DISCARD

QC
32
.A52   Allen
       Physical science
       problem guide

Physical Science Problem Guide

A. David Allen
Ricks College

Charles E. Merrill Publishing Company
A Bell & Howell Company
Columbus, Ohio

Audio-Tutorial Physical Science Series

Robert J. Foster, *Editor*

Published by
Charles E. Merrill Publishing Co.
A Bell & Howell Company
Columbus, Ohio 43216

Copyright ©, 1973 by Bell & Howell Company. All rights reserved. No part of this book may be reproduced in any form, electronic or mechanical, including photocopy, recording, or any information storage or retrieval system, without permission in writing from the publisher.

International Standard Book Number: 0-675-08939-5

Library of Congress Catalog Card Number: 73-75197

1 2 3 4 5 6 7 8 9 10 — 78 77 76 75 74 73

Printed in the United States of America

To the Instructor

This textbook and audio-cassette program has been designed to provide assistance to the general education student with the fundamentals of common mathematics problems of classical physics taught in a college physical science course. Ten different types of problems are discussed and examples and student problems are presented for each. Every effort has been made to be as complete and as clear as possible, and to give detailed help to the reader.

The audio-cassette program is designed to aid the student in reading this book and in interpreting the examples. It also provides the answers for all the problems to be worked by the student and gives detailed explanations of how those problems should be solved.

This program is not to be considered a complete rendition of any particular subject. It is designed to follow a thorough reading of the basic text that is being used in the course. Any student who is having difficulty with the mathematics problems presented in that text should be encouraged to go to this book and cassette program for detailed assistance.

To help you make the most effective use of this program in your own particular situation, the reverse side of Unit 3 has been left blank for your use in recording any special information or instructions for your own students. In order to record, place small pieces of Scotch tape over the two openings at the back of the cassette. Then, in order to ensure not erasing what you have recorded, remove the tape. You may wish to check with your audio-visual department before recording.

To the Student

In order to provide as much assistance as possible with this book, there is an individual tape available for each unit. After securing the particular tape you want, turn it on, and follow through each discussion and the examples in the book very carefully. When instructions are given to work a problem, you will be told to turn off the tape and to do the computation. After solving the problem, turn the tape back on and check your answer with the one on the tape. If your answer is incorrect, listen to the discussion of the problem which follows and, if necessary, go back to the general discussion of the material. Then rework the problem. Remember, you may rewind the tape and listen again to any section whenever you wish.

Appreciation is due to my devoted wife, Janet, for her continued support of this project, and to Miss Sharon Huffaker and Miss Renee Manwaring for their excellent typing efforts.

A. David Allen

Contents

Unit 1	Velocity, Time, Distance and Acceleration	—A simple format approach to physics problems for the non-science student	1
Unit 2	Force, Mass, Acceleration and Weight	—The mathematics of Newton's Laws in Metric and British units	17
Unit 3	Vectors, Torques	—A graphical approach to vectors using ordered pairs	29
Unit 4	Centripetal and Gravitational Forces	—An interesting way of looking at "direct and inverse" proportions using Newton's Universal Law of Gravity	39
Unit 5	Work, Power and Energy	—A good workout with the units concluding with a summary table	49
Unit 6	Pressure, Buoyancy, and Waves	—Have you ever wondered why a steel ship floats?	65
Unit 7	Temperature Scales, Gas Laws, Latent Heat	—A method for remembering temperature conversion	79
Unit 8	Magnetism, Electricity and Power	—Figure your own electrical power bill	93
Unit 9	Light and Relativity	—A careful look at the reasons for the "twins paradox"	111
Unit 10	Cathode Rays, X rays, Radioactivity	—Definitions of alpha, beta, gamma rays	123
Answers to Student's Problems			136
Index			138

1

Velocity, Time, Distance and Acceleration

FRAME 1

Velocity or *speed* can be described as the distance traveled during a particular amount of time. This means in equation form or with an equal sign: velocity = distance traveled/time required or in symbols $v = d/t$. The familiar speedometer in our automobiles has units designated as miles per hour. As we travel we often remind ourselves that the speed limit or velocity limit is 70 miles per hour or some such value. The meaning of 70 miles per hour is the fraction 70 miles traveled/in every (per) hour or 70 miles/1 hour. (As you work through this book we will write the fractions in two forms: $70 \dfrac{\text{miles}}{\text{hour}}$ and 70 miles/hour. *Both forms are equivalent.*) This is velocity and is exactly like the above equation, $v = d/t$. The numerator of the fraction is 70 miles and corresponds with the distance traveled. The denominator, in every 1 hour, corresponds to the time required.

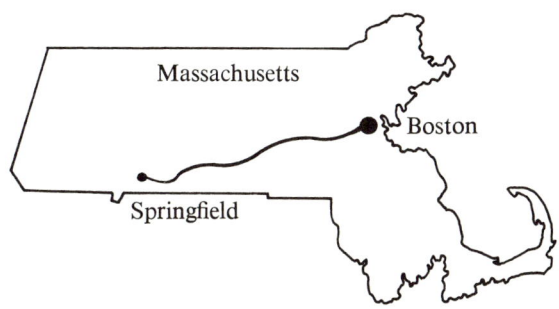

1

2 Velocity, Time, Distance and Acceleration

1. The distance from Boston to Springfield is 80 miles. The trip can be made in 2 hours. Find the average velocity for the trip.

FRAME 2

As we consider physics problems we will establish a format to help work the problems accurately and with a minimum of question. First we will list those things that are "given" and those things to "find." We will then search for a relationship between the "given" and the "find," a formula. When we find a formula that is solved for the unknown and contains all given information, we will select it. We will then replace the known quantities with symbols in the formula and then perform the indicated operations. Let's now apply this approach to problem 1.

GIVEN	FIND
Distance: $d = 80$ miles	Velocity: $v = ?$
Time traveled: $t = 2$ hours	

Since we have discussed only one relationship to this point it is the only choice. It is solved for the unknown v, and contains the given quantities d and t.

$$v = \frac{d}{t}$$

Now replace the known values for the corresponding symbols in the formula.

$$v = \frac{80 \text{ miles}}{2 \text{ hours}}$$

At this point a careful observation of the problem reveals

$$v = \frac{80}{2} \frac{\text{miles}}{\text{hours}}$$

This reduces to

$$v = \frac{40 \text{ miles}}{1 \text{ hour}} \quad \text{or} \quad 40 \text{ miles per hour}$$

Try this next problem using the format approach.

FRAME 3

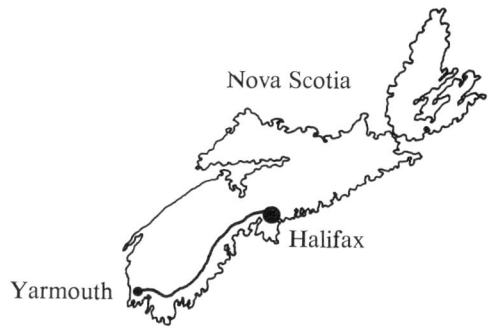

2. We made the trip in 4 hours from Halifax to Yarmouth, a distance of 200 miles. Find the average velocity for the trip.

FRAME 4

The formula $v = d/t$ is very useful not only in the form that we have considered but in another form. Consider the following manipulation. Multiply both sides of the equation by t.

$$t \cdot v = \frac{d}{t} \cdot t$$

The left side of the equation becomes t multiplied by v or tv; and in the right side the t's divide out and d remains.

$$t \cdot v = \frac{d}{\cancel{t}} \cdot \cancel{t}$$

The result

$$t \cdot v = d \quad \text{or} \quad d = v \cdot t$$

This is our first mathematical manipulation. To summarize we can say that to clear the denominator we will multiply both sides of the equation by the denominator. It is important to always perform exactly the same operation to both sides of an equation. This will maintain the statement of equality.

4   Velocity, Time, Distance and Acceleration

FRAME 5

3. A freight train will average $50 \dfrac{\text{miles}}{\text{hour}}$ or 50 miles/hour (recall our previous comment about the form of an equivalent fraction) on its run from Boise to Pocatello. The trip will take $4\frac{1}{2}$ hours. What is the distance between the two cities?
Use the ideas of the previous problem: list the "given" and "find" and proceed.

GIVEN	FIND
Velocity: $v = 50 \dfrac{\text{miles}}{\text{hours}}$	Distance between cities: $d = ?$
Time: $t = 4\frac{1}{2}$ hours	

Now as we search for a relationship our choices are $v = d/t$ or $d = v \cdot t$. Since we need to determine d and we have been given v and t, the choice is

$$d = v \cdot t$$

Now substitute the values of v and t that are given:

$$d = 50 \, \dfrac{\text{miles}}{\text{hour}} \times 4\tfrac{1}{2} \text{ hours}$$

$$d = 50 \cdot 4\tfrac{1}{2} \, \dfrac{\text{miles}}{\text{hour}} \cdot \text{hours}$$

$$d = 225 \text{ miles}$$

The hour units divided out leaving miles as the answer. Careful consideration of the units of a problem will always be a way to check answers.

FRAME 6

Try this problem:

4. A flight from Chicago to Boston takes $2\frac{1}{2}$ hours at a velocity of 400 miles/hour. What is the distance?

FRAME 7

Now to recap, our discussion started with $v = d/t$ and we transformed the equation to $d = v \cdot t$. The next step with the formula $d = v \cdot t$ is to solve for t. This can be accomplished by dividing both sides of the equation by v. Notice that the v is multiplied times the t and to eliminate any factor from one side of the equation, perform the opposite operation. The opposite operation of multiplication is division.

$$d = v \cdot t$$
$$\frac{d}{v} = \frac{v \cdot t}{v}$$
$$\frac{d}{v} = \frac{\cancel{v} \cdot t}{\cancel{v}}$$
$$\frac{d}{v} = t \quad \text{or} \quad t = \frac{d}{v}$$

FRAME 8

5. The distance from the firing point to the target is 650 feet. The average velocity of a bullet fired is 1300 ft/sec. How long will it take the bullet to reach the target?

GIVEN	FIND
Distance: $d = 650$ ft	(how long) Time: $t = ?$
Velocity: $v = 1300$ ft/sec	

6 Velocity, Time, Distance and Acceleration

As we search for a relationship we now have to make a choice from 3 possible equations.

$$v = \frac{d}{t} \qquad d = v \cdot t \qquad t = \frac{d}{v}$$

Since we are given d and v and wish to find t, the choice is

$$t = \frac{d}{v}.$$

Now make the substitution into the equation.

$$t = \frac{d}{v}$$

$$t = \frac{650 \text{ ft}}{1300 \ \dfrac{\text{ft}}{\text{sec}}}$$

$$t = \frac{650}{1300} \ \dfrac{\text{ft}}{\dfrac{\text{ft}}{\text{sec}}}$$

$$t = \frac{1}{2} \ \dfrac{\text{ft}}{\dfrac{\text{ft}}{\text{sec}}}$$

A careful manipulation of the units reveals ft/(ft/sec) means ft ÷ ft/sec which means ft·sec/ft. To divide we invert and multiply and f̶t̶·sec/f̶t̶ leaving sec. The answer is then $t = \frac{1}{2}$ sec.

FRAME 9

Try the following problem.

6. The velocity of a space station in orbit is 18,000 miles/hour. The orbit of the capsule is 27,000 miles long. How long will it take to make one complete orbit?

FRAME 10

As a car proceeds down a road and increases in velocity from 30 miles/hour to 50 miles/hour we characterize the car as *accelerating*. We

Velocity, Time, Distance and Acceleration

feel as if we are sinking back into the soft car cushions as the car increases in velocity. When the car comes to a stop we feel the opposite sensations. This is *negative acceleration* or *deceleration*.

Wheeeeee

Acceleration is due to a change in velocity during a period of time. Acceleration is also due to change in direction at constant velocity. As we round a corner at a constant velocity in our automobiles we feel a tendency to slide across the car seat. This acceleration is due to a change in direction with respect to time.

Our consideration will be to look at the problems dealing with acceleration due to a change in velocity.

FRAME 11

Acceleration expressed in equation form looks like this expression:

$$\text{acceleration} = \frac{\text{change in velocity}}{\text{time required}}$$

In symbols

$$a = \frac{V_2 - V_1}{t}$$

Where
- V_2 is the final velocity or the velocity at the end of time t.
- V_1 is the initial velocity or the velocity at the beginning of time t.
- t is the time interval.

Before considering an example, look at the units of acceleration.

$$\text{acceleration} = \frac{\text{change in velocity}}{\text{time}} =$$

8 Velocity, Time, Distance and Acceleration

(a) $\dfrac{\dfrac{\text{miles}}{\text{hour}}}{\text{seconds}}$ or (b) $\dfrac{\dfrac{\text{miles}}{\text{hour}}}{\text{hours}}$ or (v) $\dfrac{\dfrac{\text{miles}}{\text{second}}}{\text{seconds}}$

Rewriting as division problems:

(a) $\dfrac{\text{miles}}{\text{hour}} \div \text{second}$ (b) $\dfrac{\text{miles}}{\text{hour}} \div \text{hour}$ (c) $\dfrac{\text{miles}}{\text{sec}} \div \text{sec}$

$\dfrac{\text{miles}}{\text{hour}} \cdot \dfrac{1}{\text{second}}$ $\dfrac{\text{miles}}{\text{hour}} \cdot \dfrac{1}{\text{hours}}$ $\dfrac{\text{miles}}{\text{sec}} \cdot \dfrac{1}{\text{sec}}$

$\dfrac{\text{miles}}{\text{hour} \cdot \text{seconds}}$ $\dfrac{\text{miles}}{\text{hour}^2}$ $\dfrac{\text{miles}}{\text{sec}^2}$

It is important to note that acceleration is distance divided by the product of two time units such as hours·seconds, hours² or second².

FRAME 12

7. Your car approaches the freeway entrance at 30 miles/hour. You proceed onto the freeway increasing your velocity to 70 miles/hour. This takes 10 seconds. What is your acceleration?

           GIVEN   FIND

Initial velocity, at beginning of time: acceleration: $a = ?$
$V_1 = 30$ miles/hour

Final velocity, at end of time:
$V_2 = 70$ miles/hour

Time: $t = 10$ seconds

Our only available acceleration formula is

$$a = \frac{V_2 - V_1}{t}$$

$$a = \frac{70\,\frac{\text{mi}}{\text{hr}} - 30\,\frac{\text{mi}}{\text{hr}}}{10\text{ sec}} = \frac{40\,\frac{\text{mi}}{\text{hr}}}{10\text{ sec}} = \frac{40\,\frac{\text{mi}}{\text{hr}}}{10\text{ sec}} = \frac{4\,\frac{\text{mi}}{\text{hr}}}{\text{sec}}$$

The acceleration is 4 mi/hr·sec as you enter the freeway.

FRAME 13

You try this problem.

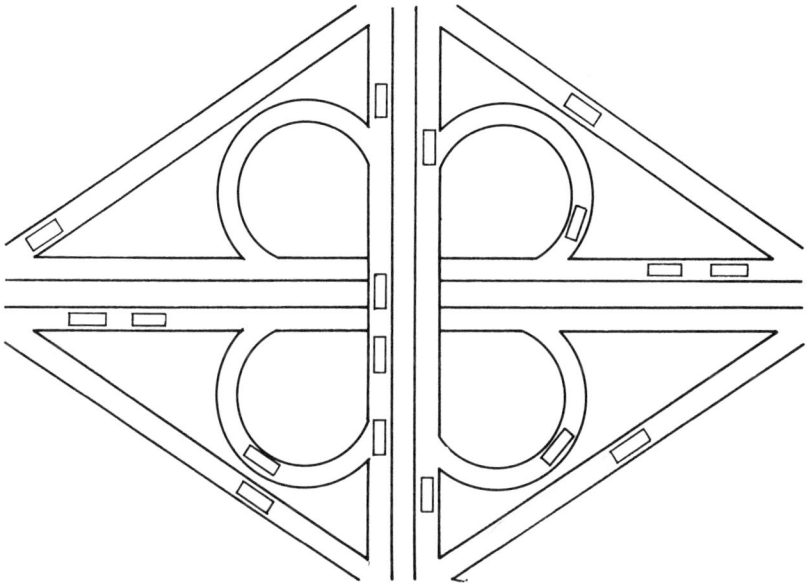

8. As you enter an expressway your velocity increases from 25 miles/hour to 60 miles/hour in 7 seconds. What is your acceleration?

FRAME 14

A problem in negative acceleration or deceleration follows.

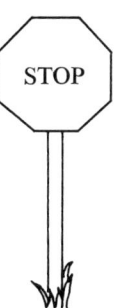

9. You approach a stop sign on a cross street at 45 miles/hour, apply the brakes and slow to a stop in 9 seconds. What is your acceleration?

10 Velocity, Time, Distance and Acceleration

GIVEN
Initial velocity: $V_1 = 45$ miles/hour
Final velocity: $V_2 = 0$ miles/hour (stopped)
Time: $t = 9$ seconds

FIND
Acceleration: $a = ?$

The acceleration formula is

$$a = \frac{V_2 - V_1}{t} = \frac{0\,\frac{mi}{hr} - 45\,\frac{mi}{hr}}{9 \text{ seconds}} = \frac{-45\,\frac{mi}{hr}}{9 \text{ seconds}} = \frac{-45}{9}\,\frac{\frac{mi}{hr}}{\text{sec}}$$

$$= \frac{-5\,\frac{mi}{hr}}{\text{sec}} = -5\,\frac{mi}{hr} \cdot \frac{1}{\text{sec}} = -5\,\frac{mi}{hr \cdot \text{sec}}$$

The negative sign is important because it notes deceleration or slowing down.

FRAME 15

Your problem:

10. A baseball catcher catches a ball traveling at 30 ft/sec and it takes .25 second ($\frac{1}{4}$ second) to slow the ball to a stop in his mitt. Find the acceleration.

FRAME 16

Each day we are conscious of the effect of the earth's acceleration on our activities. This familiar acceleration we call *gravity*. When we drop a ball, it starts with an initial velocity zero and increases in velocity till it hits the ground. As a sky diver jumps from a plane his downward velocity is initially zero and then increases as he falls. If there was no air any freely falling body would continue to increase in velocity as it fell.

Air provides a great deal of friction to falling bodies and eventually the skydiver reaches a maximum velocity.

We can calculate the velocity of any freely falling object (neglecting air friction) using the following formula:

Velocity = gravity's acceleration multiplied by time

In symbols:

$$v = g \cdot t$$

Experimentally the value for g has been determined to be 32 ft/sec/sec or 32 ft/sec². In the metric system when length is expressed in meters and time in seconds, the value for g is 9.8 meters/sec/sec or 9.8 m/sec². The fact that 32 ft is equivalent to 9.8 meters is apparent.

FRAME 17

11. How fast, in ft/sec, will a rock be falling after three seconds when dropped from a high cliff?

GIVEN	FIND
Time: $t = 3$ sec	velocity: $v = ?$

Gravity acceleration: $g = 32$ ft/sec²

We will use 32 ft/sec² not 9.8 m/sec² because the question asks for an answer in ft/sec not meters/sec.

Select the only formula discussed to this point.

$$v = g \cdot t$$

$$v = 32 \, \frac{\text{ft}}{\text{sec}^2} \times 3 \text{ sec}$$

$$v = 32 \cdot 3 \, \frac{\text{ft}}{\text{se\cancel{c}^2}} \times \text{s\cancel{e}c}$$

$$v = 96 \, \frac{\text{ft}}{\text{sec}}$$

The units of seconds on the time cancel with one of the seconds on the acceleration and units of ft/sec remain in the answer, $v = 96$ ft/sec.

FRAME 18

Your problem:

12 Velocity, Time, Distance and Acceleration

12. A skydiver jumps from a plane and reaches what velocity, in meters/sec, after 4 seconds? Neglect air friction.

FRAME 19

In both previous problems it is important to realize that the velocities with which a rock and a skydiver fall do not depend upon how much they weigh. Galileo pointed out during the sixteenth century that the weight of a falling object DOES NOT determine how fast an object will fall. Heavy objects do not fall faster than light objects if we neglect air friction.

FRAME 20

Can you solve the equation

$$v = g \cdot t \quad \text{for} \quad t?$$

Recall that to remove the factor of g we need to perform the opposite operation. Presently g is multiplied times t so we need to divide, the opposite operation.

$$v = g \cdot t$$

$$\frac{v}{g} = \frac{g \cdot t}{g}$$

$$\frac{v}{g} = \frac{\cancel{g} \cdot t}{\cancel{g}}$$

$$\frac{v}{g} = t \quad \text{or} \quad t = \frac{v}{g}$$

FRAME 21

13. A falling window washer passes your 35th floor hotel window at 80 ft/sec. How long has he been falling?

GIVEN | FIND
velocity: $v = 80$ ft/sec | How long: $t = ?$

Recall that gravity's acceleration is always 32 ft/sec² or 9.8 m/sec². Which of these two values shall we use in this problem? Notice that the velocity is given in units of ft/sec which is the British system so we must use $g = 32$ ft/sec². If the velocity had been in m/sec we would have used the other value.

Therefore

$$g = 32 \text{ ft/sec}^2$$

We have 2 formulas from which to choose.

$$v = g \cdot t \quad \text{and} \quad t = \frac{v}{g}$$

Our choice is $t = v/g$ because v and g are given and t is the unknown we are to find.

14 Velocity, Time, Distance and Acceleration

$$t = \frac{v}{g}$$

$$t = \frac{80 \frac{\text{ft}}{\text{sec}}}{32 \frac{\text{ft}}{\text{sec}^2}} = \frac{80}{32} \frac{\frac{\text{ft}}{\text{sec}}}{\frac{\text{ft}}{\text{sec}^2}} = 2.5 \frac{\frac{\text{ft}}{\text{sec}}}{\frac{\text{ft}}{\text{sec}^2}}$$

A careful consideration of the units reveals:

$$\frac{\frac{\text{ft}}{\text{sec}}}{\frac{\text{ft}}{\text{sec}^2}} = \frac{\text{ft}}{\text{sec}} \div \frac{\text{ft}}{\text{sec}^2} = \frac{\text{ft}}{\text{sec}} \cdot \frac{\text{sec}^2}{\text{ft}} = \frac{\cancel{\text{ft}}}{\cancel{\text{sec}}} \cdot \frac{\text{sec}^{\cancel{2}}}{\cancel{\text{ft}}} = \text{sec}$$

The answer is then $t = 2.5$ seconds.

FRAME 22

Your problem:

14. A skydiver jumps from a plane and begins to count the seconds as he falls. How long has he fallen when his velocity reaches 96 ft/sec?

FRAME 23

With the ideas already presented, we can derive a very important result. We will simply state it here.

The distance an object falls
 = initial velocity × time + ½ acceleration·(time²).

In symbols

$$d = V \cdot t + \tfrac{1}{2} at^2$$

If an object simply is dropped from our hand the initial velocity is 0 and the formula reduces to

$$d = \tfrac{1}{2} gt^2$$

Where a, the acceleration, is replaced with g, the acceleration of gravity. The formula $d = \frac{1}{2} gt^2$ could be written $d = gt^2/2$

FRAME 24

15. Reconsider problem 13 and its result and find how far the window washer had fallen before he came into view at your 35th floor window.

GIVEN (FROM PROBLEM 13)   FIND
Velocity: $v = 80$ ft/sec.   How far: $d = ?$

time: $t = 2.5$ sec (the answer from problem 13)

Use the formula just discussed

$$d = \tfrac{1}{2} gt^2$$

We will use $g = 32$ ft/sec² because all of our units are in ft and seconds, the British system.

$$d = \tfrac{1}{2} \cdot \left(32 \frac{\text{ft}}{\text{sec}^2}\right) \cdot (2.5 \text{ sec})^2$$

$d = \tfrac{1}{2} \cdot (32) \cdot (2.5)^2 \text{ (ft/sec}^2) \cdot \text{sec}^2 = 100 \text{ (ft/\cancel{sec}^2)} \cdot \cancel{\text{sec}^2} = 100$ ft.

Please work the following problem.

FRAME 25

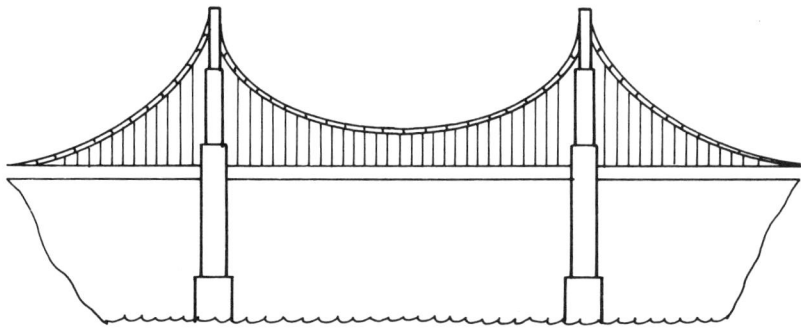

16. A rock is thrown off a high bridge. Fill in the required information in the following chart for the times given.

16 Velocity, Time, Distance and Acceleration

$t = 0$ sec	$v = 0$ ft/sec	$d = 0$ ft
$t = 1$		
$t = 2$		
$t = 3$		
$t = 4$		
$t = 4.5$		

$g = 32$ ft/sec²

The following is an example of how to work this problem for $t = 1$.

GIVEN	FIND
Time: $t = 1$ second	How fast: $v = ?$
	How far: $d = ?$

Select formulas that we have discussed.

How Fast: velocity

$v = g \cdot t$
$v = 32$ (ft/sec²) $\cdot$ 1 sec
$v = 32 \cdot 1$ (ft/sec²)·sec
$v = 32$ ft/sec

How Far: distance

$d = \frac{1}{2} gt^2$
$d = \frac{1}{2} \cdot 32$ (ft/sec²)·(1 sec)²
$d = \frac{1}{2} \cdot 32 \cdot 1$ (ft/sec²)·sec²
$d = 16$ ft

These 2 values now complete the chart to the right of $t = 1$ second. You finish the chart.

2

Force, Mass, Acceleration and Weight

FRAME 1

Sir Isaac Newton, an Englishman of the seventeenth and eighteenth centuries formulated the laws of motion into 3 concise statements.

First Law:

Every body continues in its state of rest, or of uniform motion in a straight line, unless it is acted upon by another force.

This law helps us understand why a shotput will roll over a very smooth floor for a considerable distance without stopping. If there was no friction between the floor and the shotput, and the floor was perfectly level and had no end, the shotput would never stop once put in motion. The law also helps us understand why we can jerk a table cloth out from under a plate and leave the plate undisturbed. Since the plate was at rest, not moving, the *First Law* indicates it will not be put in motion unless a

18 Force, Mass, Acceleration and Weight

force acts on it. The quantity of matter that permits this law to function is *mass*. Mass manifests itself as inertia. *Inertia* is the resistance an object has to change in its state.

FRAME 2

Second Law:

The acceleration of an object is directly proportional to the acting force and inversely proportional to the mass.

This law is summarized very neatly in the following formula:

$$\text{acceleration of object} = \frac{\text{force applied}}{\text{mass of the object}} \quad \text{in symbols, } a = \frac{f}{m}$$

FRAME 3

It is important now to consider the units or names that define the quantities of force and mass. The following chart will be worth learning and of great value in becoming familiar with these quantities.

System of Measure	Units of Mass	Units of Acceleration	Units of Force
Metric-MKS	Kilogram (Kg)	$\frac{\text{Meters}}{\text{Sec}^2}$	Newton
Metric-CGS	Gram	$\frac{\text{Centimeter}}{\text{Sec}^2}$	Dyne
British	Slug	$\frac{\text{Feet}}{\text{Sec}^2}$	Pound

The reason for emphasizing the chart at this point is that all problems that are worked in physics need to be solved consistently in one system of measure.

It is not permitted to mix units of slugs and newtons in the same problem or meters/sec^2 and pounds in the same problem.

Notice that we have not defined weight at this point. It is such a common quantity in our everyday experience. We can simply replace the *Units of*

Force, Mass, Acceleration and Weight   19

Force column in the chart with *Units of Weight*. We will consider this idea of weight later.

FRAME 4

17. What acceleration will a $\frac{1}{2}$ slug object have when pushed by a force of 3 lbs? (Neglect any frictional forces.)

GIVEN	FIND
Mass: $m = \frac{1}{2}$ slug	Acceleration: $a = ?$
Force: $f = 3$ lbs	

Using the only formula discussed

$$a = \frac{f}{m}$$

$$a = \frac{3 \text{ lbs}}{\frac{1}{2} \text{ slug}}$$

$$a = 3 \div \frac{1}{2} \frac{\text{lb}}{\text{slugs}}$$

$$a = 3 \cdot 2 \text{ (invert and multiply)} \frac{\text{lb}}{\text{slugs}}$$

$$a = 6 \frac{\text{lb}}{\text{slug}} = 6 \frac{\text{ft}}{\text{sec}^2} \quad \text{(Why these units?)}$$

FRAME 5

A careful observation of the units of this problem is now needed. To do this, we need to solve $a = f/m$ for f. The problem is to remove the m from the denominator of the right side. This is accomplished by multiplication (the opposite operation).

$$a = \frac{f}{m}$$

$$m \cdot a = \frac{f}{\not{m}} \cdot \not{m}$$

$$m \cdot a = f \quad \text{or} \quad f = ma$$

20   Force, Mass, Acceleration and Weight

The unit of force in the British system, the pound, is just a composite of mass and acceleration.

$$f = ma$$

$$\text{pounds} = \text{slug} \cdot \frac{\text{ft}}{\text{sec}^2}$$

We can say that a pound is

$$\text{slug} \cdot \frac{\text{ft}}{\text{sec}^2}$$

The answer to problem 17 was $a = 6\,\text{lb/slug}$. Making the substitution for pounds as $\text{slug} \cdot (\text{ft/sec}^2)$, the answer becomes

$$a = 6\,\frac{\text{slug} \cdot \frac{\text{ft}}{\text{sec}^2}}{\text{slug}}$$

The slug divides out, $a = 6\,([\text{slug} \cdot (\text{ft/sec}^2)]/\text{slug})$, and the answer is

$$6\,\frac{\text{ft}}{\text{sec}^2}$$

The easy way to do this type of problem is to recall that we wanted to find acceleration, and we are working a problem in the British system. Being careful only to use appropriate British units of slugs and pounds, we know acceleration is ft/sec^2 which is our answer. The above discussion is a valuable check, however.

FRAME 6

Your problem:

Force, Mass, Acceleration and Weight 21

18. A 3000 kg car is pushed with a force of 1000 newtons. What is the acceleration?

FRAME 7

In addition to solving $a = f/m$ for f as we have done, it is important to solve $a = f/m$ for m. Take the result $f = ma$ and solve for m. We will remove the factor a from the right side. This is accomplished by division (the opposite operation).

$$f = ma$$

$$\frac{f}{a} = \frac{ma}{a}$$

$$\frac{f}{a} = \frac{m \cdot \cancel{a}}{\cancel{a}}$$

$$\frac{f}{a} = m \quad \text{or} \quad m = \frac{f}{a}$$

In summary, the three forms of the equation of Newton's Second Law are:

$$a = \frac{f}{m} \qquad f = ma \qquad m = \frac{f}{a}$$

FRAME 8

19. When a .2 kg puck is accelerated 13 m/sec², what force is required?

GIVEN	FIND
Mass: $m = .2$ kg	Force: $f = ?$
Acceleration: $a = 13$ m/sec²	

Our choice of formula will be the one that is solved for f with m and a required.

$$f = ma$$

$$f = (.2 \text{ kg}) \cdot \left(\frac{13 \text{ m}}{\text{sec}^2}\right) = \frac{2.6 \text{ kg} \cdot \text{m}}{\text{sec}^2} = 2.6 \text{ newtons}$$

22 Force, Mass, Acceleration and Weight

The answer $f = 2.6 \text{ kg} \cdot \text{m/sec}^2$ is equivalent to newtons. A most important relationship to remember is that an alternate definition for a newton is $\text{kg} \cdot \text{m/sec}^2$. Can you calculate the alternate definition of a pound? Here it is.

$$f = m \cdot a$$

$$f = \text{slug} \cdot \frac{\text{ft}}{\text{sec}^2} \quad \text{(British system)}$$

$$f = \text{pound}$$

$$f = \text{kg} \cdot \frac{\text{m}}{\text{sec}^2} \quad \text{(Metric system)}$$

$$f = \text{newton}$$

FRAME 9

Newton's Second Law, $f = ma$, can be written in terms of weight by replacing the f with w and the a with g. The formula becomes:

$$w = m \cdot g$$

This new form of Newton's Second Law indicates a definition for *weight* as *the force of gravity's acceleration (attraction) upon a body*. In the morning when you step on the bathroom scales and announce that you weigh 125 pounds, you may have said, "I forced 125 pounds this morning."

Let's now perform the mathematical operations on $w = m \cdot g$ and determine the three forms.

1. $w = m \cdot g$ 2. Solve for g 3. Solve for m

$$w = m \cdot g \qquad\qquad w = m \cdot g$$

$$\frac{w}{m} = \frac{\cancel{m} \cdot g}{\cancel{m}} \qquad\qquad \frac{w}{g} = \frac{m \cdot \cancel{g}}{\cancel{g}}$$

$$\frac{w}{m} = g \quad \text{or} \quad g = \frac{w}{m} \qquad \frac{w}{g} = m \quad \text{or} \quad m = \frac{w}{g}$$

Again we divide by factors that were multiplied to solve for the desired form.

FRAME 10

Consider the following problem.

20. A man weighs 192 pounds. What is his mass?

GIVEN	FIND
Weight: $w = 192$ pounds	Mass: $m = ?$

Also known: gravity: $g = 32$ ft/sec²

The formula solved for m is selected.

$$m = \frac{w}{g}$$

$$m = \frac{192 \text{ pounds}}{\frac{32 \text{ ft}}{\text{sec}^2}}$$

$$m = \frac{6 \text{ pounds}}{\frac{\text{ft}}{\text{sec}^2}}$$

By definition of m in the British system, we know that m is measured in slugs. The verification is as follows:

$$\frac{\text{pounds}}{\frac{\text{ft}}{\text{sec}^2}} = \frac{\text{slug} \cdot \frac{\text{ft}}{\text{sec}^2}}{\frac{\text{ft}}{\text{sec}^2}} \text{ (definition of pound)} = \frac{\text{slug} \cdot \frac{\cancel{\text{ft}}}{\cancel{\text{sec}^2}}}{\frac{\cancel{\text{ft}}}{\cancel{\text{sec}^2}}} = \text{slugs}$$

The answer is $m = 6$ slugs (the mass of the man)

24 Force, Mass, Acceleration and Weight

FRAME 11

21. Determine your mass.

FRAME 12

22. A housewife purchases 16 lbs. of potatoes at the market. What mass of potatoes did she buy?

GIVEN	FIND
Weight: $w = 16$ lbs	Mass: $m = ?$
Gravity: $g = 32$ ft/sec^2	

Select the formula solved for m.

$$m = \frac{w}{g}$$

$$m = \frac{16 \text{ lbs}}{\frac{32 \text{ ft}}{\text{sec}^2}}$$

$$m = \frac{16}{32} \frac{\text{lbs}}{\frac{\text{ft}}{\text{sec}^2}} = \frac{1}{2} \frac{\text{slug} \cdot \frac{\text{ft}}{\text{sec}^2}}{\frac{\text{ft}}{\text{sec}^2}} = \frac{1}{2} \frac{\text{slug} \cdot \frac{\text{ft}}{\text{sec}^2}}{\frac{\text{ft}}{\text{sec}^2}} = \frac{1}{2} \text{ slug}$$

As we have discussed previously, pounds (lbs) are equivalent to slug·ft/sec^2 and the two fractions ft/sec^2 divide out leaving slugs as the units of the answer. The answer is $m = \frac{1}{2}$ slugs. The easy way to remember this is that the units of any problem are always the units of the unknown, and in this case the unknown is m. Thus the answer is slugs since we are working in the British system.

The parallel type problem in the metric system is to determine weight in terms of newtons and mass in kilograms if we have one of the two quantities.

FRAME 13

Your problem.

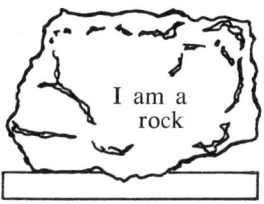

23. The mass of a rock is 8 kg. What is its weight?

FRAME 14

We now have sufficient tools to attack a problem involving several steps and using more than a single formula.

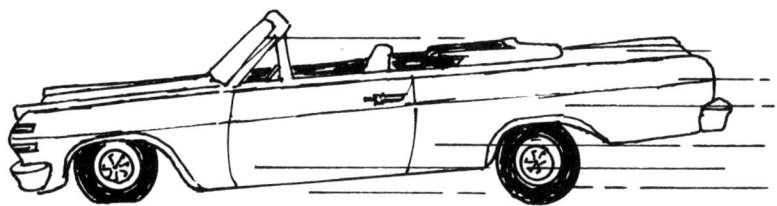

24. A 3200 lb car comes to a stop in 10 seconds from a velocity of 80 ft/sec. Find the total force the wheels had to apply to the road to stop the car.

GIVEN	FIND
Weight: $w = 3200$ lbs	Force: $f = ?$

Time: $t = 10$ seconds

Initial velocity: $v_1 = 80$ ft/sec

Final velocity: $v_2 = 0$ ft/sec (stops)

26 Force, Mass, Acceleration and Weight

Since I need to find force, my initial thought is to use the equation involving force.

$$f = ma$$

As we attempt to substitute m and a, we note that they do not have these quantities. But what is important is that we may have a way to determine m and a from the other given information.

From the given information, let's try to find a useful formula that we have discussed. As we recall all formulas to this point, we arrive at:

$$a = \frac{v_2 - v_1}{t}$$

$$a = \frac{\frac{0 \text{ ft}}{\text{sec}} - \frac{80 \text{ ft}}{\text{sec}}}{10 \text{ sec}} = \frac{\frac{-80 \text{ ft}}{\text{sec}}}{10 \text{ sec}} = \frac{-80}{10} \frac{\text{ft}}{\text{sec}} = -8 \frac{\text{ft}}{\text{sec}^2}$$

Now using the same approach, let's find m. Again we think through all the equations and select:

$$m = \frac{w}{g}$$

We have w and g given; therefore, substitute.

$$m = \frac{w}{g} = \frac{3200 \text{ lbs}}{32 \frac{\text{ft}}{\text{sec}^2}} = \frac{3200}{32} \frac{\text{lbs}}{\frac{\text{ft}}{\text{sec}^2}} = 100 \text{ slugs}$$

Recall that g in the British system is 32 ft/sec², and we used that value in this problem.

FRAME 15

The given information in this problem now becomes:

GIVEN	FIND
All previous information, plus:	Force: $f = ?$
Acceleration: $a = -8$ ft/sec² (calculated)	
Mass: $m = 100$ slugs (calculated)	

Force, Mass, Acceleration and Weight 27

We now substitute the values m and a in the formula:

$$f = ma$$

$f = (100 \text{ slugs})(-8 \text{ ft/sec}^2) = 100(-8) \text{ slug} \cdot \text{ft/sec}^2 = -800 \text{ pounds}$

A pound is slug·(ft/sec²), REMEMBER!!

The negative indicates the force is opposite to the direction of the motion or a force to slow down.

FRAME 16

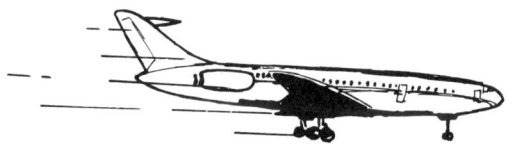

25. A 16,000 pound plane touches down at 175 ft/sec and comes to a stop in 25 seconds. Find the total force required by the wheels to stop the plane.

FRAME 17

Newton's Third Law says that for every action there is an equal and opposite reaction.

Some important principles to remember with this law are:

1. Forces always occur in pairs.

   Action — Reaction

2. The action — reaction forces never act on the same body.

Now to illustrate: As I write this sentence with a pencil on a piece of paper, let's determine the action — reaction forces and the bodies on which they act.

        Action Force: pressure of the pencil

        Reaction Force: paper pushing back

The action force is acting on the paper, and the reaction force is acting on the pencil.

FRAME 18

Consider the problem:

26. A man pulls on a rope attached to a wall with a spring balance between. The man pulls with a force of 100 pounds, and according to Newton's Third Law, the wall pulls back with 100 pounds. What then is the reading on the spring balance?

GIVEN

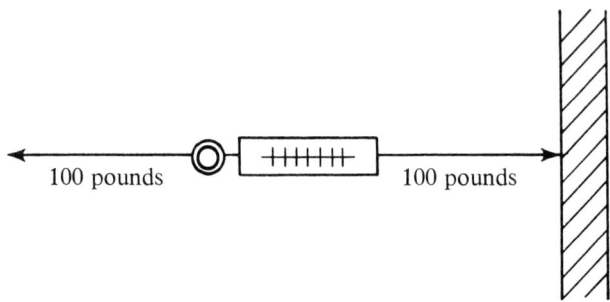

FIND

Reading on the spring balance.

Answer: 100 lbs

The action — reaction forces are equal, not additive.

FRAME 19

Your Problem.

27.

GIVEN	FIND
The following diagram with the readings on spring *A* and *C*.	Force at (*1*) and (*2*) and reading on spring *B*.

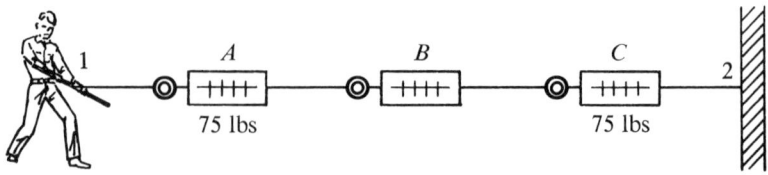

3

Vectors, Torques

FRAME 1

A very convenient way to represent many physical quantities is with the use of *vectors*. *Vectors* are quantities that have magnitude and direction. Some examples follow:

(a)
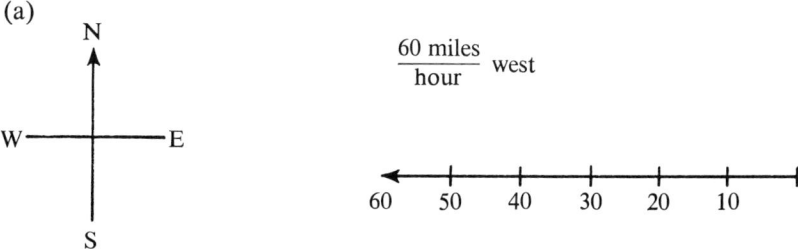

The length is 60 units, and the direction is west according to the compass.

(b) $\dfrac{10 \text{ miles}}{\text{hour}}$ southeast

Again the length is 10 units, and the direction is southeast.

FRAME 2

The following problems introduce the notion of a vector.

28. Suppose two small tractors are pulling on a stump of a tree as indicated in the diagram. Let the lengths of the chain represent the force of each tractor's pull.

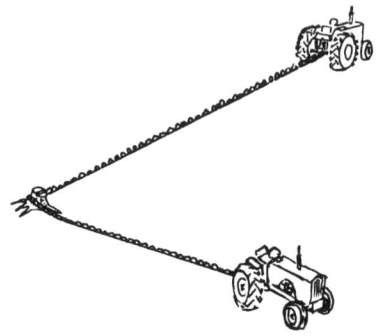

From what direction would one large tractor have to pull and with what force to equal the force of the two small tractors?

One method of solution is the parallelogram method to determine the resultant or addition of two vectors. To apply this method, consider the two vectors as sides of a parallelogram. The resultant is the length of the line from the point of applied force to the opposite vertex.

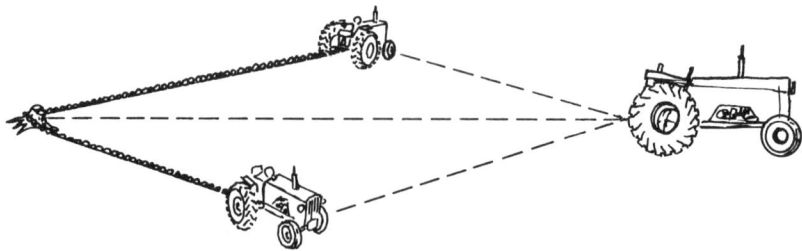

Using this method, we sketch the dashed lines to complete a parallelogram. The resultant is R. The force of the resultant is represented by the length of the line, and the direction as indicated by the position of the line. This represents the force and direction that the large tractor would have to pull to equal the two smaller tractors.

A more precise way of performing the problem can be done by superimposing a grid over the diagram, making sure the stump is at the origin (the (0,0) point).

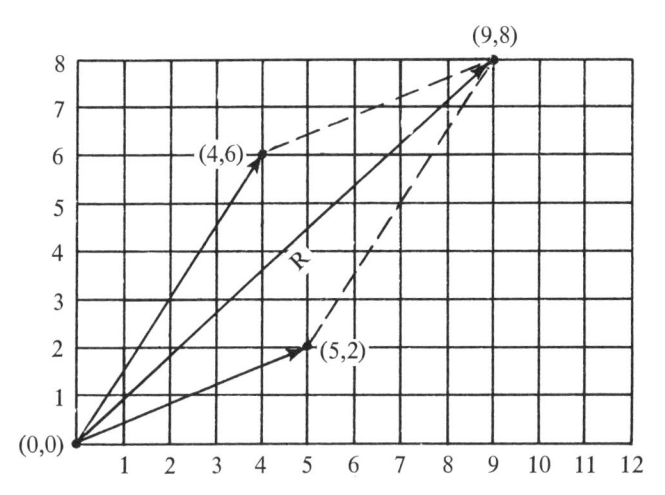

Label each arrowhead in the following way. Count right, then up and determine an ordered pair. Now simply add the vectors component-wise.

$$(4,6) + (5,2) = (9,8)$$

The resultant is from the origin to the point (9,8).

FRAME 3

29. Find the resultant of three tractors pulling on a stump.

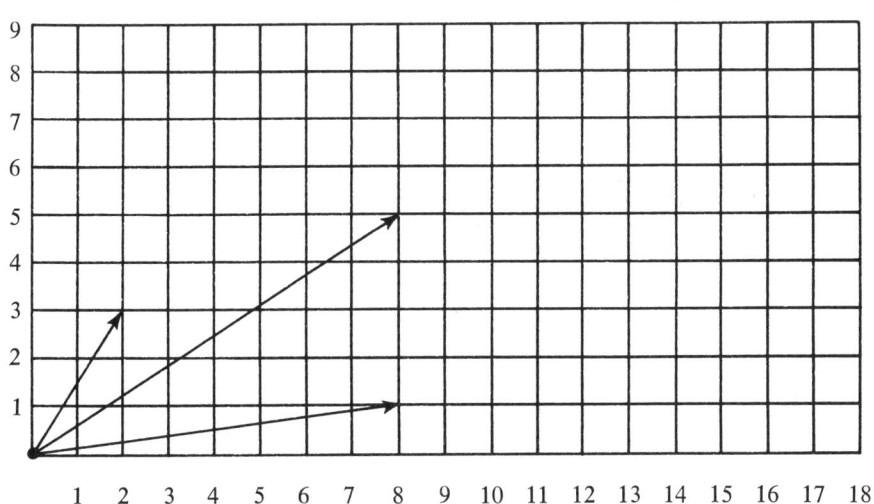

32 Vectors, Torques

FRAME 4

Consider the problem:

30. An airplane is flying due south at 80 miles/hour, and a cross wind directly to the east is blowing 60 miles/hour. Find the resultant direction and velocity of the plane.

GIVEN

$V_1 = 80$ miles/hour south

$V_2 = 60$ miles/hour east

Construct a diagram to scale.

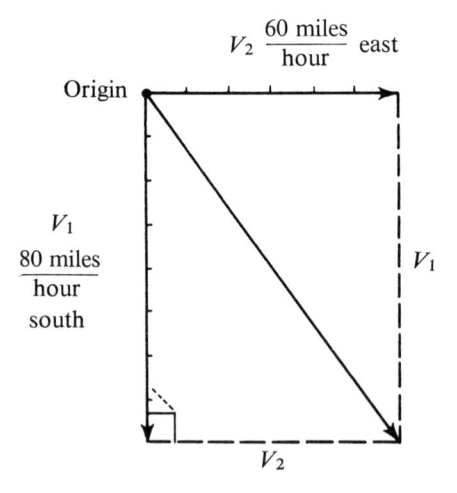

FIND

Resultant velocity and direction.

The resultant can be determined by completing the parallelogram. It is noted that in a right triangle, a triangle with one angle 90°, that the sum of the squares of the two (2) sides is equal to the square of the hypotenuse (Pythagorean's Theorem).

$$(V_1)^2 + (V_2)^2 = (\text{hypotenuse})^2$$
$$(80)^2 + (60)^2 = (\text{hypotenuse})^2$$
$$6400 + 3600 = (\text{hypotenuse})^2$$
$$10000 = (\text{hypotenuse})^2$$
$$\sqrt{10000} = \sqrt{(\text{hypotenuse})^2}$$
$$100 = \text{hypotenuse}$$

Conclusion: The resultant velocity is 100 miles/hour in a southeasterly direction.

FRAME 5

Your problem:

31. A nail is under tension from two (2) wires at 90° to each other with forces of 3 and 4 pounds. Find the resultant and the direction using the Pythagorean Theorem, and a diagram.

FRAME 6

Another type of vector problem involves resolving a vector into its horizontal and vertical components. A lady is pushing a grocery cart with force A.

Let's resolve the force she is applying to the handle into its vertical and horizontal components using the graphic method.

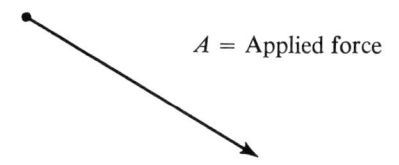

A = Applied force

Now draw vertical and horizontal lines at the origin.

34 Vectors, Torques

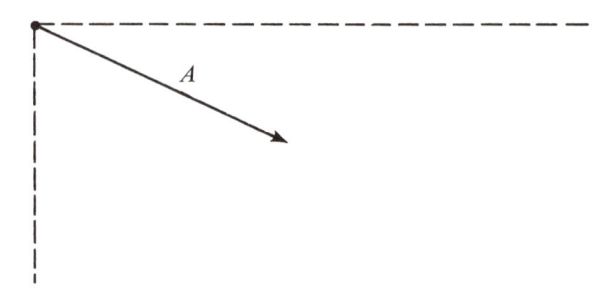

Now what size parallelogram will "box-in" the vector?

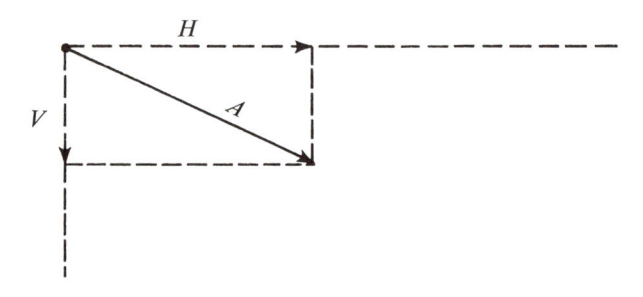

Vectors V and H represent the resolution of the vector A into its components. The H vector is what makes the cart go forward, and the V vector is a hindrance to forward movement because it just puts force on the rear wheels and this force is translated to friction.

The most economical way to push a grocery cart would be so that all the force applied is in the direction of motion. Perhaps an uncomfortable way to push a grocery cart, however.

FRAME 7

As a man pulls a heavy log with the rope attached as the diagram indicates, the resolution of the vector F is as follows:

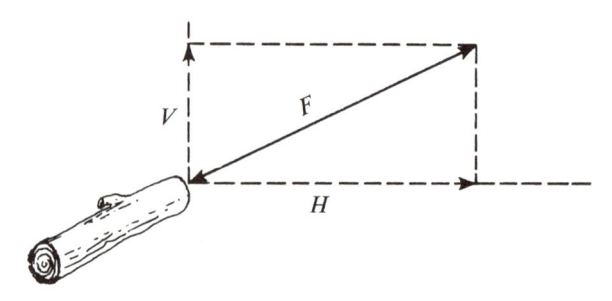

The vector V tends to lift, and the vector H drags the log.

FRAME 8

Your problem.

32. A large tractor is moving across a hillside as illustrated in the diagram. Resolve its weight vector into the H (horizontal) and V (vertical) components.

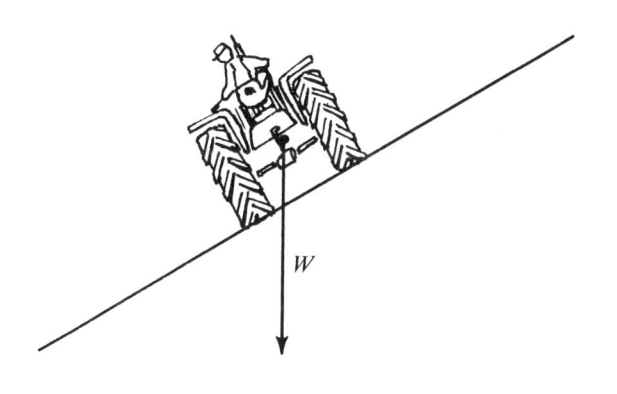

FRAME 9

Torque:

Suppose we have a teeter-totter which is off-center as illustrated, with 2 parallel forces acting at A and B, with 4 feet of length from the pivot to point B and 3 feet from the pivot to point A that has a 12 lb weight attached.

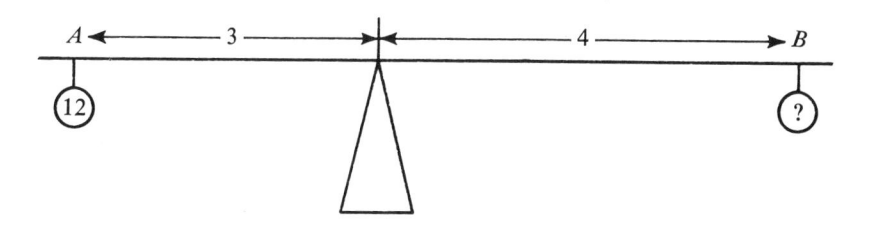

What is the force required at B to keep the board level? The *moment of force* or *torque* is the product of the lever arm length times the force.

36 Vectors, Torques

For the left side: For the right side:

$$\text{Force} \times \text{length} = \text{Force} \times \text{length}$$

$$12 \text{ lbs} \times 3 \text{ feet} = x(\text{unknown}) \times 4 \text{ feet}$$

$$36 \text{ ft·lbs} = 4x \cdot \text{feet}$$

$$\frac{36 \text{ ft·lbs}}{4 \cdot \text{ft}} = \frac{4x \text{ feet}}{4 \text{ ft}} \quad \text{Divide both sides by 4 feet}$$

$$9 \text{ lbs} = x$$

The moment of force or its torque equals the product of the force times the length of the lever arm.

FRAME 10

Consider this problem.

33. A 12-foot plank is supported at both ends by two supports, and a man weighing 200 pounds stands 4 feet from one end. What is the force on each support?

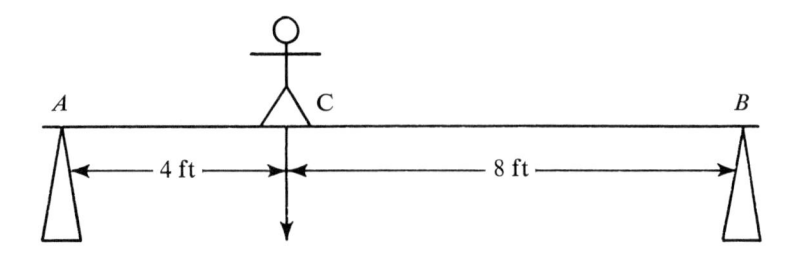

To analyze this problem, suppose a pivot point is at B.

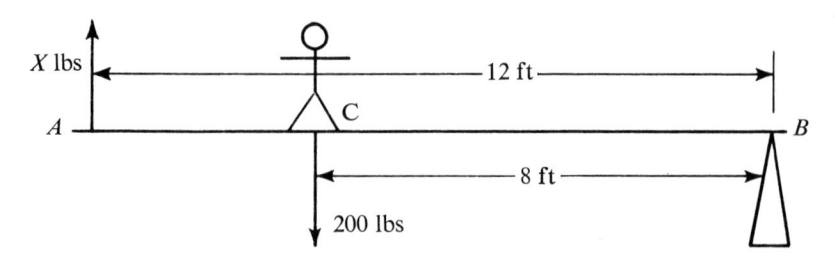

The question resolves to what is the force up A.

Vectors, Torques 37

$$\begin{aligned}
\text{torque up} &= \text{torque down} \\
\text{at } A &\qquad \text{at } C \\
(12 \text{ feet})(x \text{ lbs}) &= (8 \text{ feet})(200 \text{ lbs}) \\
12x \text{ feet} &= 1600 \text{ ft·lbs} \\
\frac{12x}{12} &= \frac{1600 \text{ ft·lbs}}{12 \text{ ft}} \\
x &= 133\tfrac{1}{3} \text{ lbs}
\end{aligned}$$

Now suppose our pivot is now at A.

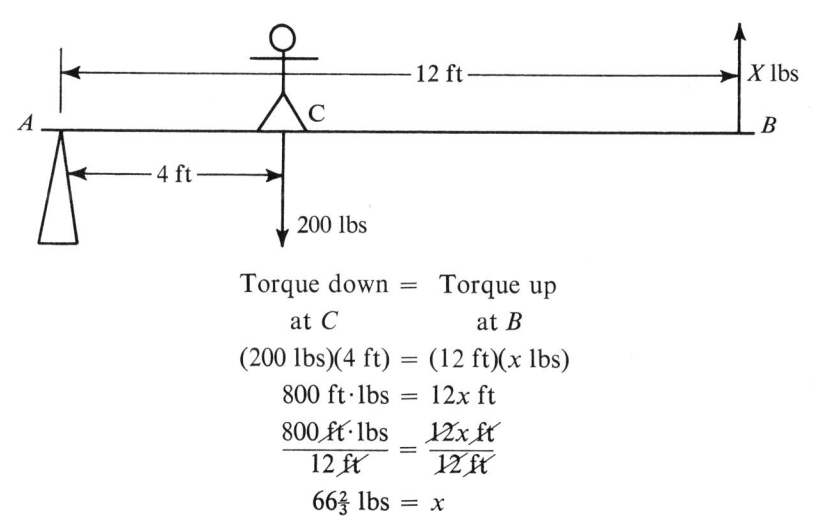

$$\begin{aligned}
\text{Torque down} &= \text{Torque up} \\
\text{at } C &\qquad \text{at } B \\
(200 \text{ lbs})(4 \text{ ft}) &= (12 \text{ ft})(x \text{ lbs}) \\
800 \text{ ft·lbs} &= 12x \text{ ft} \\
\frac{800 \text{ ft·lbs}}{12 \text{ ft}} &= \frac{12x \text{ ft}}{12 \text{ ft}} \\
66\tfrac{2}{3} \text{ lbs} &= x
\end{aligned}$$

FRAME 11

Your problem.

34. See the following diagram:

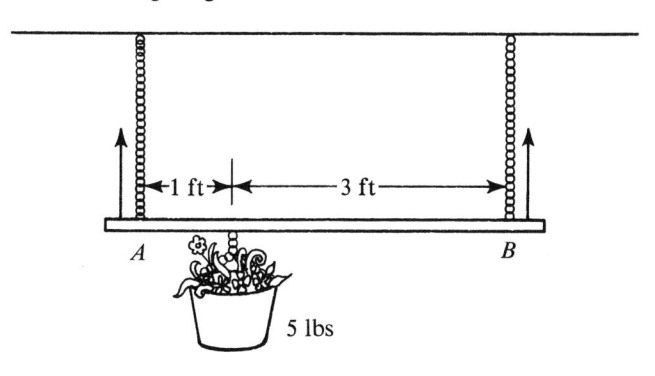

Find the forces at A and B.

4

Centripetal and Gravitational Forces

FRAME 1

Small boys often find a piece of string, tie an object on the end of the string and then proceed to whirl it about their heads in a circular motion. The circular motion of that whirling object is of some interest in our discussion. Let's analyze the motion of the object on the string as it is whirled about in a circular path.

Perhaps the first question we can ask is: Is the boy pulling on the string? To answer this question, let's suppose he lets go. Will the object continue in the circular path? The experienced answer is no. Obviously then, he is applying some force. This force being applied to the string toward the center is called the *centripetal force*.

The centripetal force can be calculated using a variation of Newton's Second Law.

$$F_c = ma$$

where a is the centripetal acceleration.

40 Centripetal and Gravitational Forces

A calculation that we shall not do here indicates that centripetal acceleration is $a = v^2/r$ and substituting this quantity into our formula for a becomes $F_c = mv^2/r$

where
$\begin{cases} F_c \text{ is the centripetal force on the string} \\ m \text{ is the mass of the object} \\ v \text{ is the velocity} \\ r \text{ is the radius of the circular path.} \end{cases}$

Again as a reminder as we work problems in physics, it will be important not to mix units from the metric and British systems in the same problem.

FRAME 2

Let's consider a problem:

35. A 1 kg ball is rotated on a chain 2 meters in length with a velocity of 13 m/sec. Find the centripetal force on the chain.

GIVEN   FIND

Mass: $m = 1$ kg   Centripetal force: $F_c = ?$

Radius: $r = 2$ meters

Velocity: $v = 13$ m/sec

Using the formula discussed to this point

$$F_c = \frac{mv^2}{r}$$

$$F_c = \frac{1 \text{ kg} \cdot \left(\frac{13 \text{ m}}{\text{sec}}\right)^2}{(2 \text{ meters})}$$

$$F_c = \frac{(1)(169)}{2} = 84\tfrac{1}{2}$$

What units will F_c be in, since we are working force units in the metric system? It will be in newtons.

Centripetal and Gravitational Forces 41

As a check, let's look at the units.

$$F_c = \frac{kg \cdot \frac{m^2}{sec^2}}{m} = \frac{kg \cdot m}{sec^2} = \text{newton}$$

They are the same units we get from $F = ma$ which is the basis of the definition of force.

FRAME 3

Your problem:

36. A discus thrower whirls an 8 lb ($\frac{1}{4}$ slug) discus around and lets it fly. If the radius of the path of the discus is 4 feet and the velocity is 20 ft/sec, what is the centripetal force on his arm?

FRAME 4

Now consider some basic manipulation of the formula and solve for m, r, and v.

The basic equation:

$$F_c = \frac{mv^2}{r}$$

Multiply both sides by r:

$$r \cdot F_c = \frac{mv^2}{r} \cdot r$$

$$r \cdot F_c = m \cdot v^2$$

42 Centripetal and Gravitational Forces

m:	r:	v:
$r \cdot F_c = m \cdot v^2$	$r \cdot F_c = m \cdot v^2$	$r \cdot F_c = m \cdot v^2$
divide both sides by v^2	divide both sides by F_c	divide both sides by m
$\dfrac{r \cdot F_c}{v^2} = \dfrac{m \cdot \cancel{v^2}}{\cancel{v^2}}$	$\dfrac{r \cdot \cancel{F_c}}{\cancel{F_c}} = \dfrac{m \cdot v^2}{F_c}$	$\dfrac{r \cdot F_c}{m} = \dfrac{\cancel{m} \cdot v^2}{\cancel{m}}$
divide out v^2	divide out F_c	divide out m
$\dfrac{r \cdot F_c}{v^2} = m$	$r = \dfrac{m \cdot v^2}{F_c}$	$\dfrac{r \cdot F_c}{m} = v^2$
		take the square root of both sides
		$\sqrt{\dfrac{r \cdot F_c}{m}} = \sqrt{v^2}$
		$\sqrt{\dfrac{r \cdot F_c}{m}} = v$

In summary, the 4 forms of the centripetal equation are:

$$F_c = \frac{mv^2}{r} \qquad r = \frac{mv^2}{F_c} \qquad m = \frac{r \cdot F_c}{v^2} \qquad v = \sqrt{\frac{r \cdot F_c}{m}}$$

FRAME 5

37. A bucket of rocks containing 1 slug of rocks is swung around in a circle of radius 2 feet. A force of 50 pounds is needed to keep it revolving. What is the velocity?

GIVEN

Mass: $m = 1$ slug
Radius: $r = 2$ feet
Centripetal force: $F_c = 50$ lbs

FIND

Velocity: $v = ?$

Look through the above equations for one solved for v with the given as their required substitute.

$$v = \sqrt{\frac{r \cdot F_c}{m}} = \sqrt{\frac{2 \text{ ft} \cdot 50 \text{ lbs}}{1 \text{ slug}}}$$

$$v = \sqrt{100}$$

Units will be ft/sec since we are working in the British system.

$$v = \frac{10 \text{ ft}}{\text{sec}}$$

A check on the units can be as follows:

$$v = \sqrt{\frac{\text{ft} \cdot \text{lbs}}{\text{slug}}} = \sqrt{\frac{\text{ft} \cdot \text{slug} \cdot \frac{\text{ft}}{\text{sec}^2}}{\text{slug}}} = \sqrt{\frac{\text{ft}^2}{\text{sec}^2}} = \frac{\text{ft}}{\text{sec}}$$

Remember! lbs are slug·ft/sec² and substituted in.

FRAME 6

38. What is the mass of a ball on a 2-meter string moving in a circular path at 8 m/sec with a force of 64 newtons on the string?

FRAME 7

Sir Isaac Newton's observation of the apple falling gave him inspiration to formulate the *Universal Law of Gravitation.*

44 Centripetal and Gravitational Forces

Universal Law of Gravitation

The law states that every object in the universe attracts every other object in the universe with a force that is directly proportional to the product of their masses and inversely proportional to the square of distance between their centers.

Again: $\dfrac{\text{gravity attraction}}{\text{between 2 bodies}} \propto \dfrac{(\text{mass one}) \times (\text{mass other})}{(\text{distance between their centers})^2}$

In symbols:

$$F \propto \frac{m_1 \cdot m_2}{d^2}$$

$\propto$ *is a proportionality symbol* and can be replaced with an $=$ sign if we include a constant of proportionality.

FRAME 8

This is our first encounter with the idea of a proportionality constant so let's explain it a little further.

Suppose we have a quarter and a penny.

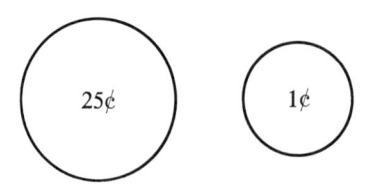

We can say that the value of one quarter is directly proportional to the value of one penny and write the expression.

n quarters $\propto n$ pennies   where n is the number of coins

Now replace the proportionality symbol with an equal sign and a proportionality constant (k). What will it be? What number will equate the value of a penny with the value of a quarter? The choice seems to be 25. Let's try it.

$$n \text{ quarters} = k \cdot n \text{ pennies}$$

$$k = \text{proportionality constant}$$

Centripetal and Gravitational Forces 45

Now let's try an example. Suppose I have 7 quarters and 7 pennies.

$$7 \text{ quarters} = (25)(7 \text{ pennies}) \quad \text{where } n = 7$$
$$\$1.75 = \$1.75$$

In the same way, we can rewrite Newton's Universal Law of Gravitation with an equal sign and a proportionality constant.

$$F = k \frac{m_1 \cdot m_2}{d^2}$$

Where F is the gravitational attraction force between two objects of masses, m_1 and m_2 are at a distance of d.

Remember:

If we want the units of force in newtons, we will need to use the metric gravitational constant — masses in kilograms and distances in meters. If we want the force in pounds, we will need to use the British gravitational constant — masses in slugs and distance in feet.

At this point, rather than to work several problems involving very large numbers of mass and distance, I believe it will be much more instructive if we explain this idea of "inversely proportional to the square of the distance" and the "product of the masses."

FRAME 9

Let's consider this problem.

39. Suppose we have 2 planets a distance of 1__ (units of distance not important) apart, with masses of 1__ (units of mass not important), and suppose the gravitational constant is 1. Compare the force between the two planets when the distance is increased to 3__ (units of distance not important). Assume for this problem $k = 1$ (the proportionality constant).

Situation 1

GIVEN	FIND
Mass 1: $m_1 = 1$	Force: $F = ?$
Mass 2: $m_2 = 1$	
Gravitational constant: $k = 1$	
Distance: $d = 1$	

46 Centripetal and Gravitational Forces

Formula:

$$F = k\frac{m_1 m_2}{d^2}$$

$$F = \frac{(1)(1)(1)}{(1)^2} = 1$$

Situation 2

GIVEN	FIND
Mass 1: $m_1 = 1$	Force: $F = ?$
Mass 2: $m_2 = 1$	
Gravitational constant: $k = 1$	
Distance: $d = 3$	

Formula:

$$F = k\frac{m_1 m_2}{d^2}$$

$$F = \frac{(1)(1)(1)}{(3)^2} = \frac{1}{9}$$

Comparison: When the planets are separated by three times, the gravitational force went from 1 to 1/9 or to 1/9 of the original value.

FRAME 10

Your problem:

40. Suppose 2 planets of mass 1 at a distance of 2 units are moved 1 unit closer. Consider the gravitational constant to be 1 and compare the 2 forces.

Centripetal and Gravitational Forces

FRAME 11

Now let's take a look at a similar type of problem, but this time vary the mass.

41. Suppose the following situations exist somewhere in the universe.

GIVEN

Situation 1:

$m_1 = 1$ ———— $d = 1$ ———— $m_2 = 1$

Situation 2:

$m_1 = 1$ ———— $d = 1$ ———— $m_2 = 2$

FIND

Comparison between the 2 forces of gravity in the 2 situations.

Let $k = 1$

Situation 1:
$$F = k\frac{m_1 m_2}{d^2}$$

$$F = \frac{(1)(1)(1)}{(1)^2} = 1$$

Situation 2:
$$F = k\frac{m_1 m_2}{d^2}$$

$$F = \frac{(1)(1)(2)}{(1)^2} = 2$$

Conclusion: By keeping all things equal except for increasing the mass to double the first value, the force is increased to double the original amount.

Centripetal and Gravitational Forces

FRAME 12

Your problem:

42.

GIVEN	FIND
The following diagram with information as indicated.	Compare the force in both situations. Let $k = 1$

Situation 1

$m_1 = 1$ ←— $d = 2$ —→ $m_2 = 2$

Situation 2

$m_1 = 2$ ←— $d = 1$ —→ $m_2 = 2$

FRAME 13

The intent of the foregoing problems has been to help you gain some feeling for the formula that Newton proposed, not to make detailed calculations of actual forces. Such a calculation for the gravitational force between the moon and the earth would be:

$$F = k \frac{\text{(mass of earth in kilograms)(mass of moon in kilograms)}}{\text{(distance in meters between centers)}^2}$$

$$F = 6.67 \times 10^{-11} \times \frac{(5.98 \times 10^{24})(7.36 \times 10^{22})}{(3.84 \times 10^8)^2}$$

$$F = \frac{(6.67)(5.98)(7.36)}{(3.84)^2} \times \frac{(10^{-11})(10^{24})(10^{22})}{(10^8)^2}$$

$$F = 1.99 \times 10^{20} \text{ newtons}$$

This force is manifest on the earth in at least one form that we are aware of — the tide.

5

Work, Power, and Energy

FRAME 1

An experience common to all of us is work. Those experiences that we define as work vary from diligent study at school to driving a car 600 miles in one day to digging a ditch. Of these experiences, some will qualify as mechanical work or work as we define it in physics, and others will not.

Work is defined as the force applied to an object multiplied by the distance it moves.

$$\text{work} = \text{force applied} \times \text{distance moved}$$

In symbols:

$$w = f \cdot d$$

This definition implies that the force has to be in the same direction as the displacement or movement. A man pushing a crate across a warehouse floor is performing work. Notice that the force is in the direction of the

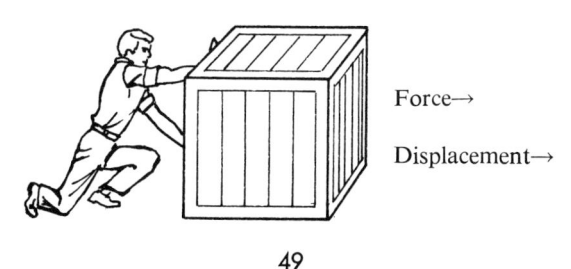

Force→

Displacement→

displacement. But a farmer walking at a constant velocity with a bucket of milk is not performing any work because the force applied to the bucket is at right angles to the displacement.

Note: This example did not say that it didn't take work to get the bucket moving.

If the force is not in the direction of the displacement, we would work a vector problem resolving the vector into its components.

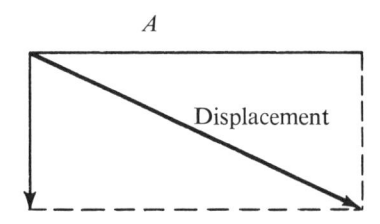

Vector A is in the direction of displacement, and its numerical value is used in the formula:

$$w = f \cdot d$$

We will consider in all of the examples that the force is applied in the direction of the displacement.

FRAME 2

Let's consider an example.

43. A man in a warehouse pushed a crate with a force of 75 pounds a distance of 25 feet. How much work did he perform?

GIVEN   FIND

Force: $f = 75$ lbs   Work: $w = ?$
Distance: $d = 25$ ft

Formula:

$$w = f \cdot d$$
$$w = (75 \text{ lbs})(25 \text{ ft}) = 1875 \text{ ft} \cdot \text{lbs}$$

1875 ft·lbs work was performed.

FRAME 3

The units of ft·lbs are the units of work in the British system. In the metric system:

$$w = f \cdot d$$
$$w = (\text{newtons})(\text{meters})$$
$$w = \text{newtons} \cdot \text{meters} = \frac{\text{kg} \cdot \text{m}}{\text{sec}^2} \cdot (\text{meters}) = \frac{\text{kg} \cdot \text{m}^2}{\text{sec}^2} = \text{joule}$$

In the CGS system:

$$w = f \cdot d$$
$$w = \text{dynes} \cdot \text{centimeters}$$
$$w = \text{dynes} \cdot \text{centimeters} = \text{erg}$$

Can you perform the mathematical manipulations and solve the equation for f and d?

$$\text{work} = \text{force} \cdot \text{distance}$$
$$w = f \cdot d$$

Solve for f and then for d.

Solve for f:

$$\frac{w}{d} = \frac{f \cdot \cancel{d}}{\cancel{d}}$$
$$\frac{w}{d} = f \quad \text{or} \quad f = \frac{w}{d}$$

Solve for d:

$$\frac{w}{f} = \frac{\cancel{f} \cdot d}{\cancel{f}}$$
$$\frac{w}{f} = d \quad \text{or} \quad d = \frac{w}{f}$$

In summary:

$$w = f \cdot d \quad f = \frac{w}{d} \quad d = \frac{w}{f}$$

52 Work, Power, and Energy

FRAME 4

Your problem:

44. A small boy drags his toy with a force of 650 dynes a distance of 150 centimeters. What is the work performed?

FRAME 5

Let's look at this problem:

45. A man performs 750 joules of work on a bucket of plaster as he lifts it with a force of 200 newtons. How high did he lift it?

GIVEN	FIND
Work: $w = 750$ joules	Distance: $d = ?$
Force: $f = 200$ newtons	

As I search for a relationship, I need to find a formula solved for d with w and f to be substituted.

$$d = \frac{w}{f} = \frac{750 \text{ joules}}{200 \text{ newtons}} = \frac{750 \text{ newtons} \cdot \text{meters}}{200 \text{ newtons}} = 3.75 \text{ meters}$$

Recall in this problem that a joule is a newton·meter.

FRAME 6

Your problem:

46. If 500 ft·lbs of work is performed on a crate to drag it 75 ft, what force is needed to move the crate?

Work, Power, and Energy 53

FRAME 7

A quantity closely associated to work is power. Power is defined as the amount of work performed in a specific amount of time.

If you perform 1000 ft·lbs of work in 5 seconds, and I perform the same amount of work in 10 seconds, the power you expend is twice as much. Therefore:

$$\text{Power} = \frac{\text{work}}{\text{time}}$$

In symbols:

$$p = \frac{w}{t}$$

Considering each of the systems of measure,

British:
$$p = \frac{w}{t} = \frac{\text{ft·lbs}}{\text{sec}}$$

When (550 ft·lbs)/sec of power is expended, we call this quantity 1 horsepower (hp).

$$1 \text{ hp} = \frac{550 \text{ ft·lbs}}{\text{sec}}$$

MKS:

$$p = \frac{w}{t} = \frac{\text{joules}}{\text{sec}} = \text{watt}$$

$1 \frac{\text{joule}}{\text{sec}}$ is defined as 1 watt.

CGS:

$$p = \frac{w}{t} = \frac{\text{erg}}{\text{sec}} = \frac{\text{erg}}{\text{sec}} \quad \text{(no special unit)}$$

As with other formulas, we need to see $p = w/t$ solved for t and w.

54 Work, Power, and Energy

Solve for w:

$$t \cdot p = \frac{w}{\cancel{t}} \cancel{t}$$

$$t \cdot p = w \quad \text{or} \quad w = p \cdot t$$

Solve for t:

$$\frac{w}{p} = \frac{\cancel{p} \cdot t}{\cancel{p}}$$

$$\frac{w}{p} = t \quad \text{or} \quad t = \frac{w}{p}$$

FRAME 8

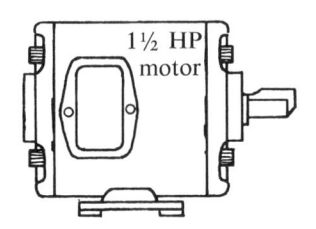

47. If a motor performs 825 ft·lbs/sec of work, what horsepower is it producing?

GIVEN | FIND
Power: $p = 825$ ft·lb/sec | Equivalent horsepower

Also known: The factor that 1 horsepower (hp) = 550 ft·lb/sec.

If 1 hp = 550 ft·lb/sec, how much hp is 825 ft·lb/sec?

$$\frac{825 \frac{\cancel{\text{ft·lb}}}{\cancel{\text{sec}}}}{550 \frac{\cancel{\text{ft·lb}}}{\cancel{\text{sec}}}} = 1\tfrac{1}{2} \text{ hp}$$
$$\overline{1 \text{ hp}}$$

FRAME 9

Your problem:

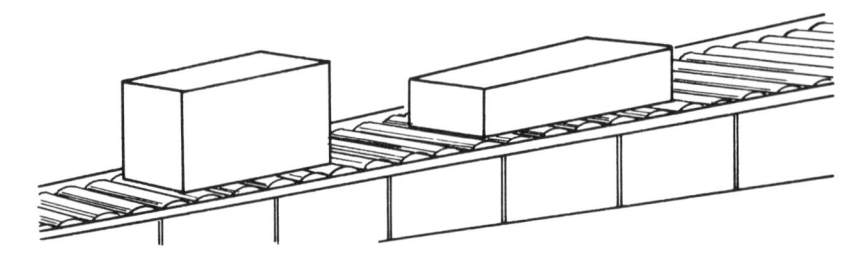

48. A conveyor is performing 412.5 ft·lb/sec of work. What is its hp?

FRAME 10

49. What is the power in horsepower (hp) of 1650 ft·lbs of work performed in 3 seconds?

GIVEN	FIND
Work: $w = 1650$ ft·lbs	Power: $p = ?$
Time: $t = 3$ seconds	

$$p = \frac{w}{t} = \frac{1650 \text{ ft·lb}}{3 \text{ sec}} = 550 \frac{\text{ft·lb}}{\text{sec}}$$

By our definition, 550 ft·lb/sec is 1 hp.

Answer: 1 hp

FRAME 11

Your problem:

50. What is the power in hp of 2200 ft·lbs of work performed in 2 seconds?

FRAME 12

51. A man can carry a 100 pound sack of feed up 22 feet (vertical distance) of stairway in 8 seconds. What horsepower will this require?

Work, Power, and Energy

GIVEN　　　　　　　　　　　　　FIND

Weight: $w = 100$ lbs — *Recall* this　Power: $p = ?$
could be called force:
$$f = 100 \text{ lbs}$$
Distance: $d = 22$ ft
Time: $t = 8$ sec

$$p = \frac{w}{t}$$

Note that I do not have the work, but the formula for work is

$$w = f \cdot d$$
$$p = \frac{f \cdot d}{t}$$

by substitution

$$p = \frac{(100 \text{ lbs})(22 \text{ ft})}{8 \text{ sec}} = \frac{2200}{8} = 275 \frac{\text{ft} \cdot \text{lbs}}{\text{sec}}$$

or 275 is what part of 550?

Answer: ½ horsepower

FRAME 13

Your problem:

52. A boy weighing 110 lbs can run up a flight of stairs (25 ft vertical distance) in 10 seconds. What horsepower will he develop?

FRAME 14

Work, Power, and Energy

Energy is the ability to do work. We can observe the manifestations of energy in many places. As a carpenter drives nails, there is obviously energy in the hammer that is derived from the strength of his arm. This energy is ultimately a chemical energy derived from the intake of nutritional food that derives its nutritional value from the biological building mechanisms that are driven by the sun.

Our consideration will be the understanding of mechanical energy as it is manifest in potential and kinetic energy.

Potential energy is the energy of position.

Kinetic energy is the energy of motion.

Now let's take a look at potential energy. We should be able to derive the mathematical formula using the basic definition. The definition implies that the potential energy of a heavy rock resting on a table is greater than that of a light rock. This would be obvious if we pushed each off the table onto the floor and observed the dents in the floor. Our conclusion is that potential energy depends upon the rock's weight or, remembering that weight equals $m \cdot g$, upon mass and gravity's acceleration.

Now let's consider another aspect of the problem. If I had 2 rocks of the same mass, placed one on the table and the other on a chair and rolled them off and observed the dents, which would be deeper? The obvious answer is the one made by the rock which fell off the table.

We can now conclude that energy of position is dependent upon:

1. mass
2. gravity's acceleration
3. height above a reference level (the floor is my reference level)

Using P.E. as a symbol for potential energy, the following formula is derived:

$$\text{P.E.} = \text{mass} \cdot \text{gravity's acceleration} \cdot \text{height}$$

or

$$\text{P.E.} = m \cdot g \cdot h$$

FRAME 15

The units are of interest.

58 Work, Power, and Energy

In the MKS system:

$$\text{P.E.} = m \cdot g \cdot h$$

$$\text{P.E.} = (\text{kg})\left(\frac{m}{\sec^2}\right)(\text{meters})$$

$$= (\text{newton})(\text{meter})$$

$$= \text{joule} \quad \text{(by our previous definition)}$$

The British system:

$$\text{P.E.} = (\text{slug})\left(\frac{\text{ft}}{\sec^2}\right)(\text{ft})$$

$$= (\text{lb})(\text{ft})$$

$$= \text{ft} \cdot \text{lb}$$

The CGS system:

$$\text{P.E.} = (\text{gm})\left(\frac{\text{cm}}{\sec^2}\right)(\text{cm})$$

$$= \text{dyne} \cdot \text{cm}$$

$$= \text{erg}$$

It is very important to realize that the energy as we are discussing it is just another form of work and can be converted into work.

FRAME 16

A problem:

53. Find the potential energy of a 7 kg rock on a table 1.5 meters high.

GIVEN	FIND
Mass: $m = 7$ kg	P.E. = ?
Height: $h = 1.5$ m	

We know gravity's acceleration: $g = 9.8$ m/sec²

$$\text{P.E.} = m \cdot g \cdot h$$

$$\text{P.E.} = (7 \text{ kg})\left(9.8 \frac{m}{\sec^2}\right)(1.5 \text{ meters})$$

$$\text{P.E.} = 102.9 \text{ newtons}$$

FRAME 17

Your problem:

54. What is the potential energy of a sledge hammer weighing 16 pounds raised to a height of 5 feet?

FRAME 18

At this point, you should be able to solve P.E. $= m \cdot g \cdot h$ for m and h and solve problems using the other forms of the equation. Solve for m and h.

FRAME 19

The other type of mechanical energy that we will consider is Kinetic Energy or energy of motion. The formula is not as obvious as the P.E. formula and so it is stated:

$$\text{K.E.} = \tfrac{1}{2}mv^2$$

Kinetic energy of a moving body is dependent upon the multiplication of $\frac{1}{2}$ times the mass multiplied by the square of the velocity.

Consider the Units

MKS System:

$$\begin{aligned}\text{K.E.} &= \text{kg} \cdot \left(\frac{\text{m}}{\text{sec}}\right)^2 \\ &= \text{kg} \cdot \frac{\text{m}}{\text{sec}^2} \cdot \text{m} \\ &= \text{newton} \cdot \text{m} \\ &= \text{joule}\end{aligned}$$

Work, Power, and Energy

British System:

$$\text{K.E.} = \text{slug} \cdot \left(\frac{\text{ft}}{\text{sec}}\right)^2$$

$$= \text{slug} \cdot \frac{\text{ft}}{\text{sec}^2} \cdot \text{ft}$$

$$= \text{lb} \cdot \text{ft}$$

$$= \text{ft} \cdot \text{lb}$$

CGS System:

$$\text{K.E.} = \text{gm} \cdot \left(\frac{\text{cm}}{\text{sec}}\right)^2$$

$$= \text{gm} \cdot \frac{\text{cm}}{\text{sec}^2} \cdot \text{cm}$$

$$= \text{dyne} \cdot \text{cm}$$

$$= \text{erg}$$

FRAME 20

55. Find the K.E. of a girl on a bicycle going 22 ft/sec down the road. The bike and the girl weigh 160 pounds.

GIVEN	FIND
Velocity: $v = 22$ ft/sec	K.E. = ?
Weight: $w = 160$ lbs	

Answer: K.E. = $\tfrac{1}{2}mv^2$

We do not have the mass. Can we determine the mass from the information that we do have?

Work, Power, and Energy 61

Remember: $w = m \cdot g$ and $m = w/g$

Therefore:

$$m = \frac{160 \text{ lbs}}{32 \frac{\text{ft}}{\text{sec}^2}}$$

$$m = 5 \text{ slug}$$

$$\text{K.E.} = \tfrac{1}{2}mv^2$$

$$\text{K.E.} = \tfrac{1}{2} \cdot 5 \text{ slug} \cdot \left(22 \frac{\text{ft}}{\text{sec}}\right)^2$$

$$\text{K.E.} = \tfrac{1}{2} \cdot 5 \cdot 484$$

$$\text{K.E.} = \tfrac{1}{2} \cdot 5 \cdot 484$$

$$\text{K.E.} = 1210 \text{ ft} \cdot \text{lbs}$$

A careful effort will point out that the units of (slug)(ft/sec)² equals ft·lbs.

FRAME 21

56. You can push your 3200 lb car 1 ft/sec on a level driveway. Find the K.E. of the moving car.

FRAME 22

The formula K.E. = $\tfrac{1}{2}mv^2$ can now be solved for m and v.

The formula can alternately be written as:

$$\text{K.E.} = \frac{mv^2}{2} \text{ which is the same as K.E.} = \tfrac{1}{2}mv^2$$

Solve for m:

Multiply both sides by 2

$$2(\text{K.E.}) = \frac{mv^2}{2} \cdot 2$$

$$2(\text{K.E.}) = mv^2$$

62 Work, Power, and Energy

Now divide both sides by v^2

$$\frac{2(K.E.)}{v^2} = \frac{m\cancel{v^2}}{\cancel{v^2}}$$

$$\frac{2(K.E.)}{v^2} = m$$

Now solve for v: Taking the formula $2(K.E.) = mv^2$

Divide by m

$$2(K.E.) = mv^2$$

$$\frac{2(K.E.)}{m} = \frac{\cancel{m}v^2}{\cancel{m}}$$

$$\frac{2(K.E.)}{m} = v^2$$

Take the square root of both sides

$$\sqrt{\frac{2(K.E.)}{m}} = \sqrt{v^2}$$

$$\sqrt{\frac{2(K.E.)}{m}} = v$$

The forms of the Kinetic energy formula are:

$$K.E. = \tfrac{1}{2}mv^2 \qquad m = \frac{2(K.E.)}{v^2} \qquad v = \sqrt{\frac{2(K.E.)}{m}}$$

FRAME 23

57. Find the velocity of a 1 kg mass with a K.E. of 5000 joules.

GIVEN   FIND

Mass: $m = 1$ kg   Velocity: $v = ?$

Kinetic Energy: $K.E. = 5000$ joules

The only formula of the 3 above that is solved for the unknown is:

$$v = \sqrt{\frac{2 \cdot (K.E.)}{m}}$$

$$v = \sqrt{\frac{2 \cdot 5000}{1}}$$

$$v = \sqrt{10,000}$$

$$v = 100 \; \frac{\cancel{ft} \; m}{sec}$$

FRAME 24

Your problem:

58. What is the mass of a body moving 20 ft/sec with a K.E. of 800 ft·lbs?

Work, Power, and Energy

This is perhaps a good place to summarize all the units that we have considered to this time.

System	Distance	Mass	Time	Velocity	Acceleration	Force (weight)	Work	Power
MKS	meter	kilogram	second	$\dfrac{m}{sec}$	$\dfrac{m}{sec^2}$	newton $\left(\dfrac{kg\text{-}m}{sec^2}\right)$	newton-m (joule)	$\dfrac{joule}{sec}$ (watt)
CGS	centimeter	gram	second	$\dfrac{cm}{sec}$	$\dfrac{cm}{sec^2}$	dyne $\left(\dfrac{g\text{-}cm}{sec^2}\right)$	dyne-cm (erg)	$\dfrac{erg}{sec}$
British	foot	slug	second	$\dfrac{ft}{sec}$	$\dfrac{ft}{sec^2}$	pound $\left(\dfrac{slug\cdot ft}{sec^2}\right)$	ft·lb	$\dfrac{ft\cdot lb}{sec}$

$$550\,\frac{ft\cdot lb}{sec} = 1\ hp$$

6

Pressure, Buoyancy, and Waves

FRAME 1

When 4 small boys lie down on the lawn one on top of the other, the boy on the bottom of the pile is subjected to considerable pressure. It is this quantity that we now wish to discuss.

By definition, pressure is defined as force per unit area.

In symbols:

$$\text{pressure} = \frac{\text{force}}{\text{area}}$$

$$p = \frac{f}{a}$$

If the 3 top boys weigh an average of 75 pounds each and the area of the boy on the bottom is 300 square inches or in², the average pressure is:

$$\text{pressure} = \frac{3 \cdot 75}{300 \text{ in}^2} = \frac{225 \text{ lbs}}{300 \text{ in}^2} = \frac{9}{12} = \frac{3 \text{ lbs}}{4 \text{ in}^2}$$

FRAME 2

59. A woman weighing (forcing) 108 pounds with a heel on her shoes $\frac{3}{4}$ inches square balances on one heel. What is the pressure against the floor in in^2?

GIVEN

Weight: $w = 108$ (or force)

Dimension of shoe:
 $d = \frac{3}{4}$ in. square heel

FIND

Pressure: $p = ?$ in lbs/in²

$$\text{pressure} = \frac{\text{force}}{\text{area}} = \frac{108}{?}$$

The calculation of area is length times width or:

$$\text{area} = \text{length} \cdot \text{width}$$

$$\text{area} = l \cdot w$$

And in this problem:

$$\text{area} = \frac{3}{4} \text{ in} \cdot \frac{3}{4} \text{ in} = \frac{9}{16} \text{ in}^2$$

Therefore:

$$\text{pressure} = \frac{108 \text{ lbs}}{\frac{9}{16} \text{ in}^2} = \frac{108 \text{ lbs}}{\frac{9}{16} \text{ in}^2}$$

Invert and multiply:

$$108 \div \frac{9}{16} = 108 \cdot \frac{16}{9} = 192 \frac{\text{lbs}}{\text{in}^2}$$

That hurts if it's on your toe!

FRAME 3

Your problem:

60. A 180 lb man wears shoes with area of 20 in² each. How much pressure does he exert on the ground in lbs/in²? *Remember* he has 2 shoes.

FRAME 4

61. A large-track tractor weighs 48,000 pounds and has 2 tracks that are each 8 feet 4 inches long and 2 feet wide. Find the pressure in lbs/in² that the 2 tracks exert on the ground.

GIVEN	FIND
Weight or force: $w = 48,000$ lbs	Pressure: $p = ?$

Dimension of 2 tracks in inches

Area: $a = 100$ in $\cdot$ 24 in $\cdot$ 2(tracks)
$= 4800$ in²

$$p = \frac{48,000 \text{ lbs}}{4800 \text{ in}^2} = \frac{10 \text{ lbs}}{\text{in}^2}$$

FRAME 5

Can you perform the manipulation with the formula:

$$\text{pressure} = \frac{\text{force}}{\text{area}}$$

and solve for force and area?

68 Pressure, Buoyancy, and Waves

Solve for f: $\qquad$ Solve for a:

$$a \cdot p = \frac{f}{\cancel{a}} \cdot \cancel{a} \qquad \frac{f}{p} = \frac{\cancel{p} \cdot a}{\cancel{p}}$$

$a \cdot p = f$ or $f = p \cdot a \qquad \frac{f}{p} = a$ or $a = \frac{f}{p}$

force = pressure·area $\qquad$ area = $\dfrac{\text{force}}{\text{pressure}}$

FRAME 6

Your problem:

62. A 3000-lb automobile exerts pressure of 15 lbs/in² on the rubber that touches the road. How much area of rubber touches the road?

FRAME 7

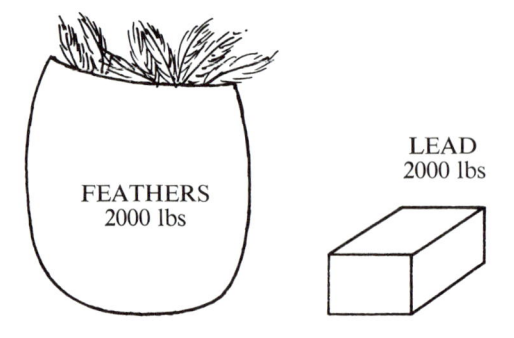

FEATHERS
2000 lbs

LEAD
2000 lbs

We are all aware of the old joke that questions — Which weighs more, a ton of feathers or a ton of lead? The observant person notes that a ton of anything is equivalent to a ton of anything else, but then what is the quantity that tends to make this question confusing? It is density. Which is less dense — a ton of feathers or a ton of lead? The answer is, of course, feathers.

By definition, density is weight (force) per unit volume.*

---

*We will not consider the other type of density, mass density, in this discussion.

Pressure, Buoyancy, and Waves 69

In symbols:

$$\text{density} = \frac{\text{weight}}{\text{volume}}$$

$$d = \frac{w}{v}$$

Solve for *w*:

$$v \cdot d = \frac{w}{\cancel{v}} \cdot \cancel{v}$$

$$w = d \cdot v$$

Solve for *v*:

$$\frac{v \cdot \cancel{d}}{\cancel{d}} = \frac{w}{d}$$

$$v = \frac{w}{d}$$

Some familiar substances and their densities are listed.

Gasoline	42.00 $\frac{\text{lbs}}{\text{ft}^3}$
Water	62.00 $\frac{\text{lbs}}{\text{ft}^3}$
Concrete	140.00 $\frac{\text{lbs}}{\text{ft}^3}$
Gold	1200.00 $\frac{\text{lbs}}{\text{ft}^3}$
Air	.08 $\frac{\text{lbs}}{\text{ft}^3}$
Ice	58.00 $\frac{\text{lbs}}{\text{ft}^3}$
Oak Wood	45.00 $\frac{\text{lbs}}{\text{ft}^3}$
Steel	480.00 $\frac{\text{lbs}}{\text{ft}^3}$

We will need to understand the computation of volume for the following problems. Volume is defined as length · width · height.

In symbols: volume = $l \cdot w \cdot h$

FRAME 8

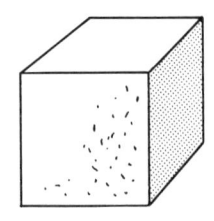

63. A square block of concrete 2 ft on each side weighs 1120 lbs. What is the density of the concrete?

GIVEN FIND

Weight: $w = 1120$ lbs   Density: $d = ?$

Dimensions: $l = 2$ ft

$w = 2$ ft

$h = 2$ ft

Volume: $v = l \cdot w \cdot h$

$v = 2 \text{ ft} \cdot 2 \text{ ft} \cdot 2 \text{ ft} = 8 \text{ ft}^3$

$$d = \frac{w}{v}$$

$$d = \frac{1120 \text{ lbs}}{8 \text{ ft}^3}$$

$$d = 140 \frac{\text{lbs}}{\text{ft}^3}$$

Notice that this value agrees with the table.

FRAME 9

Your problem:

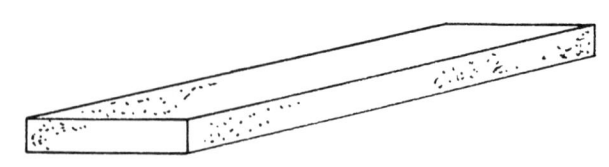

64. A piece of foam rubber is 2 feet in length, 1 foot wide, and 3 inches thick. It weighs 5 lbs. Find the density in lbs/ft³. *Remember* that 3 inches = ¼ foot.

FRAME 10

A careful observation of the densities listed in the foregoing discussion will show that the density of ice is 58 lbs/ft³ and water is 62 lbs/ft³. Our experience tells us that ice will float. The general rule is that less dense fluids and solids will float when immersed in more dense fluids. Will a solid piece of oak float in a pan of gasoline? Check the chart!

The obvious question arises, why then will steel ships float in water? The answer comes in the understanding of Archimedes' Principle that states that an object is buoyed up by a force equal to the weight of the displaced fluid.

Buoyant force = weight of displaced fluids

Suppose a battle ship were frozen in an Arctic strait, and the ice was so deep that it froze the entire hull including the bow. Suppose a great sky-hook came along and snatched the ship out of the ice, and all that remained was a frozen hole in the ice. The *weight of water* it would take to completely fill the hole is equal to the weight of the ship.

FRAME 11

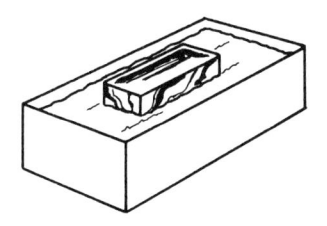

72 Pressure, Buoyancy, and Waves

65. An empty metal box (1 ft · 1 ft · 1 ft) floats in a tub of water. The box sinks into the water 4 inches. How much does the box weigh?

GIVEN	FIND
Volume: $v = l \cdot w \cdot h$	Weight of the box: $w = ?$

$$v = 1 \cdot 1 \cdot \tfrac{1}{3}$$
$$= \tfrac{1}{3} \text{ ft}^3 \text{ displaced}$$

Density of water: $d = 62 \text{ lbs/ft}^3$

Recall that according to Archimedes' Principle, the weight of displaced water equals buoyant force.

$$w = d \cdot v$$
$$\text{weight} = 62 \frac{\text{lbs}}{\text{ft}^3} \cdot \frac{1}{3} \text{ ft}^3 = 20\tfrac{2}{3} \text{ lbs}$$

Since it displaced $20\tfrac{2}{3}$ lbs, it must weigh $20\tfrac{2}{3}$ lbs.

FRAME 12

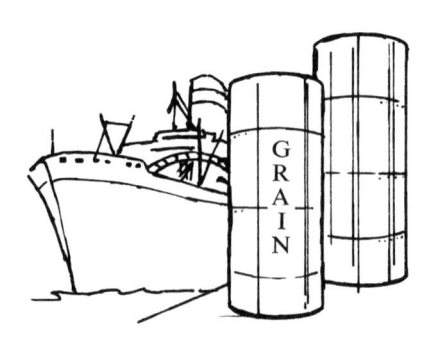

66. A large ocean liner docks at a pier and begins loading grain. As the grain is loaded, the ship sinks deeper and deeper into the water. It displaces an additional 100,000 cubic feet during loading. Find the weight of grain loaded.

FRAME 13

67. A piece of material, volume 1 cubic foot, sinks into gasoline 6 inches. Its dimensions are 1 ft long, 1 ft wide, 1 ft high. How much does it weigh?

Pressure, Buoyancy, and Waves 73

GIVEN	FIND
Volume: $v = 1$ ft³	Weight of material: $w = ?$

Volume displaced: $v = \frac{1}{2}$ ft³

Density of gasoline: $d = 42$ lbs/ft³

Recall again that a material displaces a weight of liquid equal to its weight.

$$w = d \cdot v$$
$$w = 42 \cdot \tfrac{1}{2} = 21 \text{ lbs}$$

Since it displaced a weight of 21 lbs, then it must weigh 21 lbs.

FRAME 14

68. A piece of oakwood 1 cubic foot in volume is put in a tub of water. How far does it sink? The density of oakwood is 45 lbs/ft³.

FRAME 15

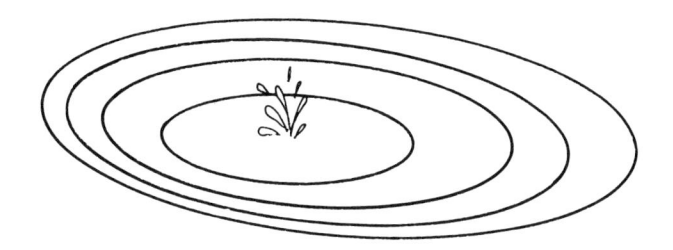

When a rock is dropped into a quiet pool, a series of concentric waves are emitted from the point of disturbance. We have come to call them water waves. If the disturbance is changed to a severe earthquake and the pool to an ocean, the damaging waves that reach the shore are tsunamis or seismic waves. We are also familiar with sound waves that can convey intelligent conversations, beautiful music, or the racket of a motorbike muffler. The colors that we observe to be different such as green and red are due to a wave phenomenon.

A very useful representation of a wave is borrowed from trigonometry — the sine wave.

74 Pressure, Buoyancy, and Waves

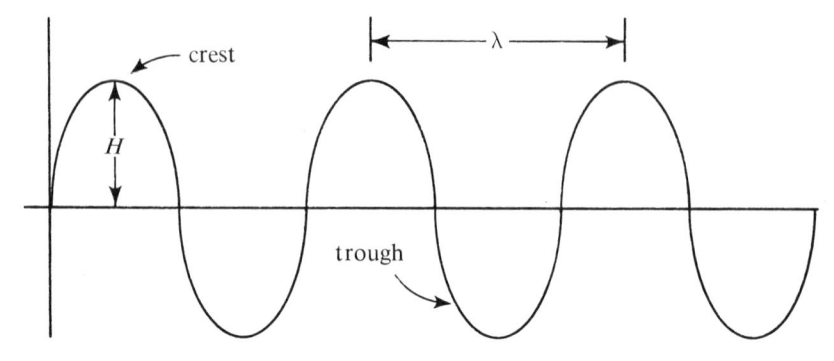

This is only a representation of a wave and not what the wave is actually like.

The amplitude (or wave height) is measured in feet or meters, the wave length (λ) is measured in feet/cycle or meters/cycle.

FRAME 16

Now think of this wave as being in motion to the right. The frequency is the number of crests, or troughs, that you observe pass your observation point every second. The units of frequency are cycles/sec, and the velocity is the speed in m/sec or ft/sec that a crest or trough moves along.

The relation between these quantities is:

$$\text{velocity} = \text{wave length} \cdot \text{frequency}$$
$$v = \lambda \cdot f$$

$$\text{MKS:} \quad \frac{m}{sec} = \frac{meter}{cycle} \cdot \frac{cycle}{sec}$$

$$\text{British:} \quad \frac{ft}{sec} = \frac{ft}{cycle} \cdot \frac{cycle}{sec}$$

$$\text{CGS:} \quad \frac{cm}{sec} = \frac{cm}{cycle} \cdot \frac{cycle}{sec}$$

Solve for f:

$$\frac{v}{\lambda} = \frac{\cancel{\lambda} \cdot f}{\cancel{\lambda}}$$

$$\frac{v}{\lambda} = f$$

or

$$f = \frac{v}{\lambda}$$

Solve for λ:

$$\frac{v}{f} = \frac{\lambda \cdot \cancel{f}}{\cancel{f}}$$

$$\frac{v}{f} = \lambda$$

or

$$\lambda = \frac{v}{f}$$

It is important to note that amplitude is not part of the formula.

Some important physical constants that we will need are the velocity of sound in different substances.

Speed of sound in:

$$\text{Air} \quad 1{,}100 \; \frac{\text{ft}}{\text{sec}}$$

$$\text{Water} \quad 4{,}800 \; \frac{\text{ft}}{\text{sec}}$$

$$\text{Steel} \quad 16{,}000 \; \frac{\text{ft}}{\text{sec}}$$

FRAME 17

Consider this problem:

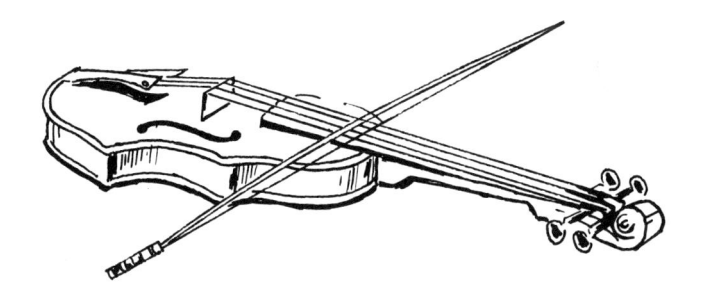

69. A violin string is plucked so that it vibrates 550 times a second (cycle/sec). What is the wavelength of the sound?

GIVEN

Frequency: $f = 550$ cycle/sec

Physical constant: Speed of sound in air — 1100 ft/sec

FIND

Wavelength: $\lambda = ?$

Answer

$$\lambda = \frac{v}{f} = \frac{1100 \; \frac{\text{ft}}{\text{sec}}}{550 \; \frac{\text{cycle}}{\text{sec}}} = 2 \; \frac{\text{ft}}{\text{cycle}}$$

FRAME 18

70. The middle C sound on the piano is approximately 4.2 feet in length. Find the frequency.

FRAME 19

71. The wavelength of red light is .0000007 meters. The velocity of light is 300,000,000 m/sec. Find the frequency.

GIVEN   FIND

Wavelength: λ = .0000007 meters   Frequency: $f = ?$

Velocity light:
v = 300,000,000 meters/sec

This problem is significant because we are working with very large and small quantities. It is easy to work with these quantities if we change them to powers of ten notation.

For numbers less than 1, move the decimal to the right until a number between 1 and 10 results. Count the decimal places moved and write that number as a negative power of 10.

$$\lambda = .0000007. = 7.0 \times 10^{-7} \text{ meters}$$

Decimal moved 7 places to the right.

For numbers greater than 10, move the decimal to the left until a number between 1 and 10 results. Count the decimal places moved and write that number as a positive power of 10.

$$v = 3.00,000,000. = 3 \times 10^8 \, \frac{m}{sec}$$

Decimal moved 8 places to the left.

Now substitute in the formula.

$$f = \frac{v}{\lambda} = \frac{3 \times 10^8 \, \frac{m}{sec}}{7 \times 10^{-7} \, \frac{meters}{cycle}}$$

$$= \frac{3}{7} \times \frac{10^8}{10^{-7}}$$

$$= .43 \times 10^{15}$$

$$= 4.3 \times 10^{-1} \times 10^{15}$$

$$\lambda = 4.3 \times 10^{14} \frac{\text{cycles}}{\text{sec}}$$

Rules for exponents tell us we can bring an exponent from the denominator and change its sign.

FRAME 20

72. Determine the wavelength of the shade of violet light with frequency 9×10^{14} cycle/sec.

Recall velocity of light — 3×10^8 m/sec

7

Temperature Scales, Gas Laws, Latent Heat

FRAME 1

As we observe substances about us we have experiences with solid materials, materials that flow which are called liquids, and the seldom noticed gases. Solids, liquids, and gases are the three common phases of matter. In our experience we seldom encounter one substance that manifests itself in all three states in common experience. One substance that we are familiar with in each of its states is water. As a solid it is ice, as a liquid it is water, and as a gas it is steam or water vapor.

One of the most significant influences on the states of matter is the effect of temperature. Let's consider four scales used to measure temperature, noticing the relationship of numerical values on each scale.

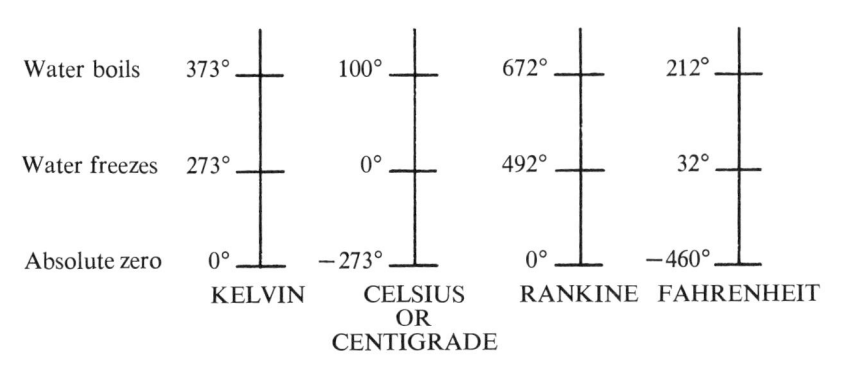

	KELVIN	CELSIUS OR CENTIGRADE	RANKINE	FAHRENHEIT
Water boils	373°	100°	672°	212°
Water freezes	273°	0°	492°	32°
Absolute zero	0°	−273°	0°	−460°

It is of interest to know that Kelvin and Centigrade degrees are the same size and that Rankine and Fahrenheit degrees are the same size but smaller than Kelvin and Centigrade. In our discussion we will be most concerned with Fahrenheit and Celsius and converting temperatures from one to the other. Also, we will need to be somewhat familiar with the Kelvin scale when working with the gas laws.

FRAME 2

Let's just look at conversion between Celsius and Fahrenheit. We could at this point consider the conversion formulas that follow:

$$T_f = \frac{9}{5} T_c + 32° \quad \text{and} \quad T_c = \frac{5}{9}(T_f - 32°)$$

However, it is easier to remember the following system than the formulas.

The Conversion System

1. Add 40 to the temperature to be changed.
2. Multiply by
    (a) 5/9 if converting from F° to C°
    (b) 9/5 if converting from C° to F°
3. Subtract 40 from the result.

FRAME 3

73. Convert 27° Celsius to Fahrenheit.

GIVEN	FIND
Temperature: $T = 27°C$	Temperature: $T = ?°F$

Step 1

$$\begin{array}{r} 27 \\ + 40 \\ \hline 67 \end{array}$$

Step 2

$$67 \times \frac{9}{5} = \frac{603}{5} = 120\frac{3}{5}$$

Temperature Scales, Gas Laws, Latent Heat   81

Step 3

$$\begin{array}{r} 120\frac{3}{5} \\ -40 \phantom{\frac{3}{5}} \\ \hline 80\frac{3}{5} \end{array}$$

Answer: $T = 80\ 3/5\ °F$ (equivalent to 27° Celsius)

FRAME 4

Your problem:

74. Convert $-48°F$ to Celsius.

FRAME 5

One other type of conversion that we should consider is from Celsius to Kelvin. This procedure is done simply by adding 273. A quick check of the four temperature scales in the preceding section will point this out.

FRAME 6

75. Change 0°C to °K.

GIVEN	FIND
Temperature: $T = 0°C$	Equivalent Temperature: $T = ?°K$

Answer:
$$\begin{array}{r} 0°C \\ +273 \\ \hline T = 273°K \end{array}$$

FRAME 7

Your problem:

76. Convert $-15°C$ to K.

FRAME 8

The consideration of temperature and the states of matter brings us to the interesting topic of *latent heat*. Our experience reminds us that as a chunk of ice in a pan on a stove is warmed, it changes states from ice to water (melts) and then from water to steam (boils). In physics it is of interest to consider the amount of heat required to perform these changes. Before we consider this, let's not forget that the reverse, the removal of heat from an object, is the function of a refrigerator. A warm bottle of soda pop can be cooled by the removal of heat.

The unit of heat in the MKS system is kilocalories; in the CGS, calories; and in the British, B.T.U. A kilocalorie is the amount of heat required to raise 1 kg of water 1°C. A calorie is the amount of heat required to raise 1 gm of water 1°C. The B.T.U. is the amount of heat required to raise 1 lb of water 1°F.

FRAME 9

To help our explanation, let's look at a graph.

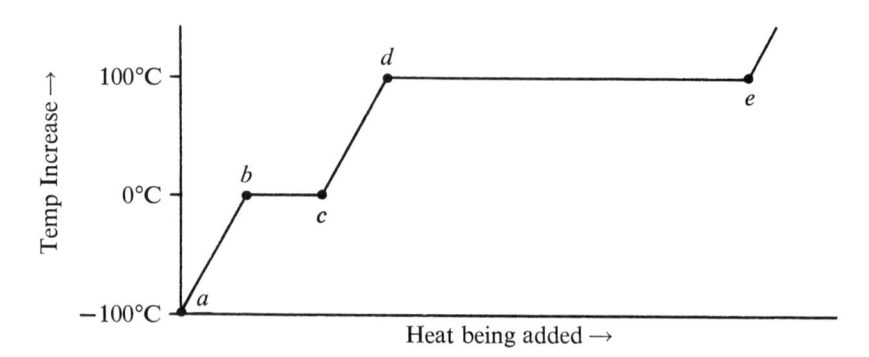

Suppose at position a we have a quantity of ice that is very cold, below freezing. As we add heat, the ice warms up to temperature b. At temperature b it starts to melt. The change from a to b requires .55 kilocalories/kg.

From b to c we continue to add heat but the ice-water mixture does not warm. The reason is because the heat being added is used to break the bonds in the ice and not to increase the temperature of the mixture. It requires 80 kilocal/kg of heat to change 1 kg ice at 0°C to 1 kg water at 0°C.

At point c all the ice melts and the liquid ice (water) now increases in temperature from c to d. It requires 1 kilocal/kg to raise water 1°C. At point d boiling commences and from d to e the liquid is now being changed to steam. Notice that the water doesn't get any hotter, but we still continue to add heat. The heat being added is changing the water to steam. It requires 540 kilocal/kg to change water to steam. At point e all the water is changed to steam, and from e to f the steam is being heated.

FRAME 10

77. How much heat does it take to change ¾ kilogram of ice at 0°C to steam at 100°C?

GIVEN	FIND
Grams of ice = ¾ kilograms	Heat required in kilocalories

One approach to this problem is to consider each of the changes in order, then total the heat required.

1. Ice 0°C to water 0°C $\quad 80 \dfrac{\text{kcal}}{\text{kg}} \times \dfrac{3}{4} \text{ kg} = 60 \text{ kcal}$

2. Water 0°C to water 100°C $\quad 1 \dfrac{\text{kcal}}{\text{kg} \cdot {}°\text{C}} \times \dfrac{3}{4} \text{ kg} \times 100°\text{C} = 75 \text{ kcal}$

3. Water 100°C to steam 100°C $\quad 540 \dfrac{\text{kcal}}{\text{kg}} \times \dfrac{3}{4} \text{ kg} = 405 \text{ kcal}$

$\qquad\qquad\qquad\qquad\qquad\qquad$ *Answer:* Total $\quad$ 540 kcal

FRAME 11

78. Work a similar problem using ½ kg of ice.

FRAME 12

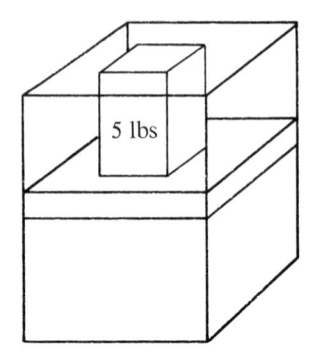

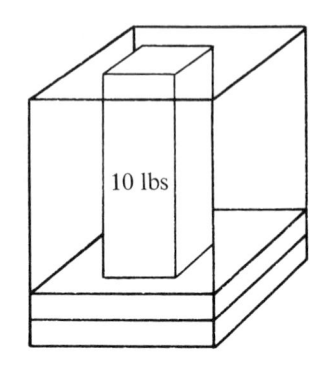

As we consider the gaseous state of matter, it is of interest to look carefully at the effect of 3 variables on a quantity of gas — volume, pressure, and temperature. As these quantities vary, they affect the other two in some way. Pressure, volume, and temperature of gas samples are related in simple formulas. In each of Boyle's, Charles's, and Gay-Lussac's formulas we will hold one of the quantities constant and develop the relationship between the other two quantities.

Boyle's Law:

If the temperature of a gas is kept constant, the volume of a quantity of gas is inversely proportional to the pressure applied. In other words, if you increase the pressure on a quantity of gas, keeping the temperature constant, the volume decreases.

That means:

$$\text{pressure} \times \text{volume} = \text{constant}$$

or

$$\text{pressure case 1} \times \text{volume case 1} = \text{pressure case 2} \times \text{volume case 2}$$

In symbols:

$$P_1 \cdot V_1 \quad = \quad P_2 \cdot V_2$$
$$\text{(Case 1)} \quad \text{(Case 2)}$$

The inversely proportional idea can be explained in this example if sets of two numbers are multiplied together and the product is 12.

For example, 1 × 12 = 12. If the first number becomes 2, what is the second? 2 × 6 = 12. If the first number becomes 3 then 3 × 4 = 12. As the first number increased, the second decreased and the numbers are said to be inversely proportional. We then can say the pairs of numbers 1 : 12, 2 : 6, 3 : 4 are inversely proportional.

FRAME 13

The units for volume and pressure are:

>volume — ft³, m³, cm³
>
>pressure — lbs/in², centimeters of mercury, atmospheres

Perhaps a comment about the units of pressure is in order. The relationship lbs/in² is familiar to you but the other two may not be.

The weight of air pressing down on the surface of the earth will hold a column of mercury in a sealed, evacuated tube 76 cm high. This pressure corresponds to 14.7 lbs/in² or 1 atmosphere.

FRAME 14

79. The pressure on 3 ft³ of gas is 2 atmospheres. If the pressure is decreased to ½ atmosphere, what is the new volume if the temperature remains constant?

GIVEN	FIND
Case 1: $V_1 = 3$ ft³	$V_2 = ?$
$P_1 = 2$ atmospheres	
Case 2: $P_2 = \frac{1}{2}$ atmosphere	

Using Boyle's Law:

$$P_1 \cdot V_1 = P_2 \cdot V_2$$
$$(2 \text{ at})(3 \text{ ft}^3) = (\tfrac{1}{2} \text{ at})(V_2)$$
$$6 = \frac{V_2}{2}$$
$$2 \cdot 6 = \frac{V_2}{\cancel{2}} \cdot \cancel{2}$$
$$12 = V_2$$
$$12 \text{ ft}^3 = V_2$$

86 Temperature Scales, Gas Laws, Latent Heat

FRAME 15

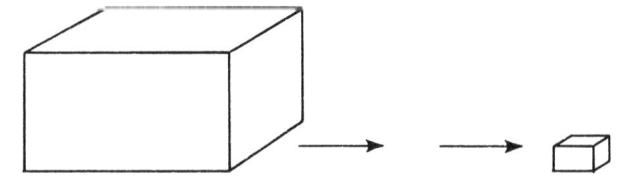

80. A quantity of gases under 1 atmosphere pressure is compressed to 5 atmospheres and 2 cubic feet. What was the original volume? Assume the temperature was held constant.

FRAME 16

Charles's Law:

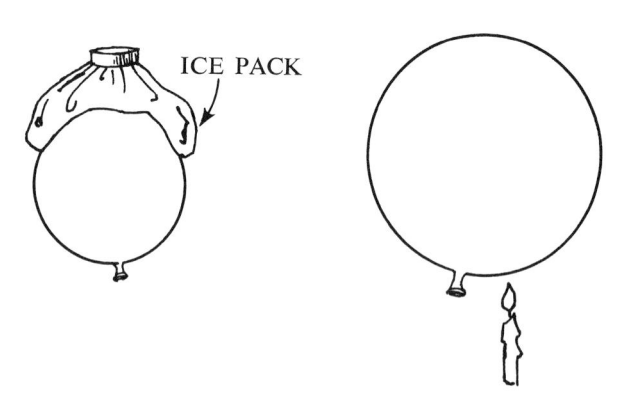

ICE PACK

If the pressure is held constant, the volume of a gas is directly proportional to the change in temperature measured in degrees Kelvin. In other words, if you heat a quantity of gas, the volume increases assuming no change in pressure.

That means:

$$\frac{\text{volume}}{\text{Temperature (Kelvin)}} = \text{constant}$$

In symbols:

(Case 1)   (Case 2)
$$\frac{V_1}{T_1} = \frac{V_2}{T_2}$$

Temperature Scales, Gas Laws, Latent Heat   87

A directly proportional statement has the following meaning. If pairs of numbers are divided and the product is 15, then $30/2 = 15$. If the numerator becomes 45, then $45/3 = 15$. If the numerator becomes 60, then $60/4 = 15$. Summary: As the numerator increases, the denominator increases to maintain the constant 15. We then say that the pairs of numbers 30 : 2, 45 : 3, 60 : 4 are directly proportional.

FRAME 17

81. A quantity of gas 50 ft³ is maintained at 27°C. The temperature is then decreased to $-73°C$. Find the new volume. The pressure remains constant.

GIVEN	FIND
Case 1: $V_1 = 50$ ft³	$V_2 = ?$

$T_1 = 27°C$ (Kelvin $= 27° + 273 = 300°$)

Case 2: $T_2 = -73°C$ (Kelvin $= -73° + 273 = 200°$)

$$\frac{V_1}{T_1} = \frac{V_2}{T_2}$$

$$\frac{50 \text{ ft}^3}{300} = \frac{V_2}{200}$$

$$200 \cdot \frac{50}{300} = \frac{V_2}{200} \cdot 200$$

$$\frac{200 \cdot 50}{300} = V_2$$

$$\frac{100}{3} = V_2$$

$$33\tfrac{1}{3} = V_2$$

FRAME 18

82. A quantity of gas, 30 ft³, is heated from 0°C to some unknown temperature. The volume increases to 60 ft³. Find the new temperature. (Pressure is held constant.)

88 Temperature Scales, Gas Laws, Latent Heat

FRAME 19

Gay-Lussac's Law

A third law, known as Gay-Lussac's Law, states that pressure of a gas is directly proportional to the Kelvin temperature assuming the volume remains constant. Another way to say this is that if you heat a quantity of gas in a fixed volume container (like a pressure cooker), the pressure will go up.

That means pressure divided by temperature is a constant.
In symbols:

$$\frac{P}{T} = k$$

or

(Case 1)   (Case 2)
$$\frac{P_1}{T_1} = \frac{P_2}{T_2}$$

FRAME 20

83. A closed vessel of gas at 1 atmosphere pressure is heated from 127°C to 327°C. What is the new pressure?

GIVEN

Case 1: Pressure — $P_1 = 1$ atmosphere

Temperature —
$T_1 = 127°C$ or
$(273 + 127) = 400°K$

FIND

Pressure Case 2: $P_2 = ?$

Case 2: Temperature —

$T_2 = 327°C$ or

$(273 + 327) = 600°K$

$$\frac{P_1}{T_1} = \frac{P_2}{T_2}$$

$$\frac{1 \text{ at}}{400°K} = \frac{P_2}{600°K}$$

$$600 \cdot \frac{1}{400} = \frac{P_2}{600} \cdot \cancel{600}$$

$$\frac{600 \cdot 1}{400} = P_2$$

Answer: $1\frac{1}{2}$ atmosphere $= P_2$

FRAME 21

84. If the pressure in a closed container at 27°C drops from 1 atmosphere to $\frac{1}{4}$ atmosphere, what must be the new temperature to create such a decrease?

FRAME 22

Of more interest to us than the individual gas laws is the Ideal Gas Law, which is just a combination of the three laws of the foregoing discussion.

Boyle's Law: Charles's Law: Gay-Lussac's Law:

$P_1V_1 = P_2V_2$ $\dfrac{V_1}{T_1} = \dfrac{V_2}{T_2}$ $\dfrac{P_1}{T_1} = \dfrac{P_2}{T_2}$

Now, noting that in all cases pressure is inversely proportional to volume and both pressure and volume are directly proportional to temperature, then

$$\frac{P_1V_1}{T_1} = \frac{P_2V_2}{T_2}$$

This equation relates all three variables assuming we use pressure in atmosphere, cm of mercury, or lbs/in^2; temperature in degrees Kelvin; and volume in customary units of cm^3, ft^3, m^3.

FRAME 23

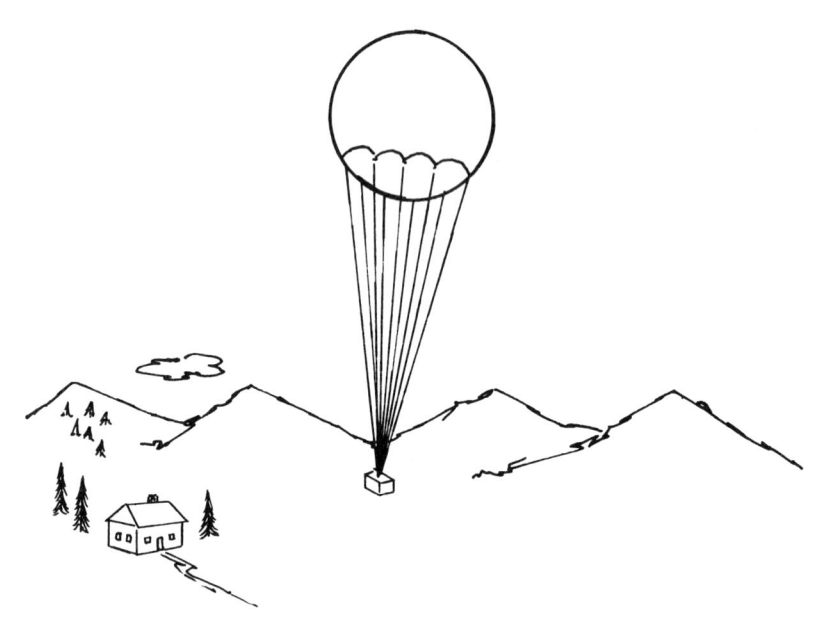

85. A 1000 ft³ balloon is readied for flight at ground level. The temperature is 27°C and the pressure is 1 atmosphere. The balloon is allowed to float high into the sky. The temperature drops to −3°C and pressure to .6 atmosphere. Find the volume of the balloon.

GIVEN   FIND

Volume: V_1 = 1000 ft³   Volume 2: V_2 = ?

Temperature 1: T_1 = 27°C

  (273 + 27° = 300°K)

Pressure 1: P_1 = 1 atmosphere

Temperature 2: T_2 = −3°C

  + 273°C

  = 270°K

Pressure 2: P_2 = .6 atmosphere

It is interesting to note that the cooling effect on the balloon will tend to reduce the volume while the reduction of pressure because of altitude

Temperature Scales, Gas Laws, Latent Heat 91

will tend to expand the volume. It will be interesting to see which of the two counteracting influences is predominant.

$$\frac{P_1 V_1}{T_1} = \frac{P_2 V_2}{T_2}$$

$$\frac{(1 \text{ atm})(1000 \text{ ft}^3)}{(300°\text{K})} = \frac{(.6 \text{ atm})(V_2)}{(270°\text{K})}$$

$$\frac{270}{.6} \times \frac{(1)(1000)}{300} = \frac{(.6)(V_2)}{270} \times \frac{270}{.6}$$

$$\frac{(270)(1)(1000)}{(.6)(300)} = V_2$$

Answer: $1500 \text{ ft}^3 = V_2$

The results indicate expansion from 1000 ft³ to 1500 ft³. Pressure reduction was the predominant influence.

FRAME 24

86. You purchased a helium balloon at the fair. The temperature is 35°C, and the volume of the balloon is 1 ft³. The balloon gets away from you and floats into the sky. It expands to 1.5 ft³, then pops. The temperature at that height is still 35°C. Find the pressure when the balloon popped. Pressure at ground level was 1 atmosphere.

8

Magnetism, Electricity, and Power

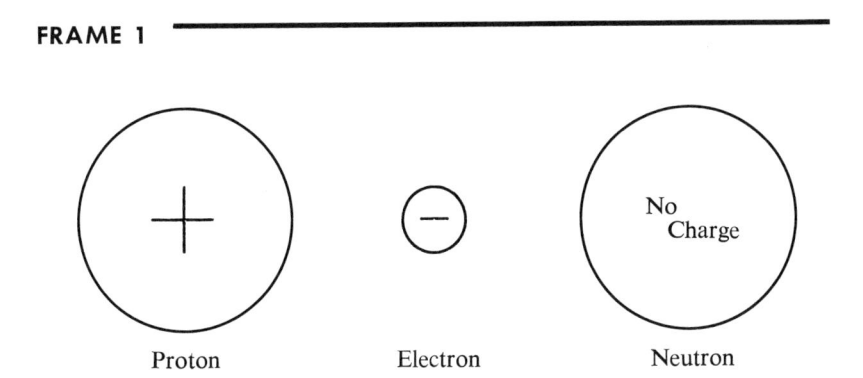

FRAME 1

+	−	No Charge
Proton	Electron	Neutron

Our discussions to this point have included the attraction of celestial bodies on the cosmic scale that we know as gravitational attraction. We discussed Newton's Law of Gravity on page 44. This force, as strong as it may seem, is very weak when compared to the forces at the other end of the spectrum in the atom. Much of physics focuses on the nature of the atom and its characteristics. Some of the basic information that is known concerns its electrical charge. We may consider all matter to be composed of atoms and all atoms to be composed of electrons, protons, and neutrons. Further subdivision of these particles is possible but not necessary for our discussion. Experimental results have concluded the electron to have a negative charge, the proton to have a positive charge, and the neutron to be neutral.

FRAME 2

The meaning of *charge can be thought of as the presence or absence of electrons*. If I am charged negatively after dragging my feet across the living room carpet, I have an *excess* of negative charge or electrons. If, on the other hand, I rub a glass rod with a silk scarf, the glass rod is charged positively or has a *deficiency* of electrons.

An interesting experiment is to suspend two balloons from threads and charge each. If both are charged negatively, they repel. If both are charged positively, they also repel. But if they are given opposite charges, they attract. The forces of attraction or repulsion are as real as the forces of gravitation.

FRAME 3

This relationship of attraction or repulsion of charge has been structured into a formula known as *Coulomb's Law*.

Coulomb's Law

The force of attraction or repulsion is directly proportional to the product of the magnitude of the charges and inversely proportional to the square of the distance.

In symbols:

Attraction or repulsion:

$$F \propto \frac{q_1 q_2}{d^2}$$

where q_1 and q_2 are the measures of the charges in units of coulombs, and d is the distance of separation of the charges in meters.

At this point let's recall the form of Newton's Universal Law of Gravitation.

$$F = k\frac{m_1 m_2}{d^2}$$

where m_1 and m_2 are the masses of the objects, and d is the distance.

The striking evidence here is that the two formulas are very similar; this helps us with the computation. Having understood the inverse square relationship and the measuring of a constant of proportionality in our discussion of gravity, we can apply these ideas to problems involving charge.

FRAME 4

Recall that with gravity we replaced the $\propto$ (proportionality symbol) with an equal sign by substituting a k or a proportionality constant. In the same way we can then rewrite Coulomb's Law as:

Attractive or repulsive:

$$F = k \cdot \frac{q_1 q_2}{d^2}$$

Where k is Coulomb's constant.

FRAME 5

Consider this problem:

87. Two charges, each of positive charge 1____ at a distance of 2____ are moved to a distance of 4____. Compare the forces in the two positions. Assume the proportional constant is 1. Notice that no units are used for this exercise.

96 Magnetism, Electricity, and Power

Situation 1

GIVEN	FIND
Charge: $q_1 = 1$	Force: $f = ?$
Charge: $q_2 = 1$	
Proportionality constant: $k = 1$	
Distance: $d = 2$	

Situation 2

GIVEN	FIND
Charge: $q_1 = 1$	Force: $f = ?$
Charge: $q_2 = 1$	
Proportionality constant: $k = 1$	
Distance: $d = 4$	

Situation 1: $f = \dfrac{(1)(1)(1)}{(2)^2} = \dfrac{1}{4}$

Situation 2: $f = \dfrac{(1)(1)(1)}{(4)^2} = \dfrac{1}{16}$

Conclusion: By moving the charge from 2 to 4, or double distance, the force diminished from $\frac{1}{4}$ to $\frac{1}{16}$ or to $\frac{1}{4}$ the original value ($\frac{1}{4}$ of $\frac{1}{4}$ is $\frac{1}{16}$).

There is one more comment that need be made and that is since the two charges were positive, the force was repulsive.

If you need assistance in this type of problem, please return to the discussion on page 46 and review it.

FRAME 6

Your problem:

88. Compare the force of 2 positive charges of charge 1 and 3, 1 unit apart, and a positive charge of 2 and a negative charge of 3, 1 unit apart.

Magnetism, Electricity, and Power 97

FRAME 7

Now let's consider the computation of an actual problem in electrostatics.

89. A positive charge of 2.5×10^{-8} coulomb is located .06 meters (6 cm) from a negative charge of 7.2×10^{-8} coulomb. Calculate the force between them. Coulomb's constant in the MKS system is 9×10^9.

GIVEN	FIND
Proportionality constant:	Force: $F = ?$
$k = 9 \times 10^9$

Charge 1: $q_1 = 2.5 \times 10^{-8}$

Charge 2: $q_2 = 7.2 \times 10^{-8}$

Distance: $d = .06$ meters or
6.0×10^{-2}

$$F = k \cdot \frac{q_1 q_2}{d^2}$$

$$F = 9 \times 10^9 \cdot \frac{(2.5 \times 10^{-8})(7.2 \times 10^{-8})}{(6.0 \times 10^{-2})^2}$$

or

$$(36 \times 10^{-4})$$

One approach to this type of problem is to separate the powers of 10.

$$F = \frac{(9)(2.5)(7.2)}{(36)} \times \frac{(10^9)(10^{-8})(10^{-8})}{(10^{-4})}$$

Now calculate each individually.

$$F = 4.5 \times 10^{-3} \text{ newtons}$$

And the force is attractive since the two charges are opposite.

FRAME 8

Your problem:

90. A positive charge of 5×10^{-8} coulomb is located 8 cm or .08 meters from another positive charge of 1.0×10^{-7} coulomb. Calculate the force.

FRAME 9

Suppose you were given all the quantities in Coulomb's Law except the distance. Could you solve Coulomb's Law for d?

$$F = k \cdot \frac{q_1 q_2}{d^2}$$

Clear the denominator.

$$d^2 F = \frac{k q_1 q_2}{\cancel{d^2}} \cdot \cancel{d^2}$$

Divide both sides by F.

$$\frac{d^2 \cancel{F}}{\cancel{F}} = \frac{k q_1 q_2}{F}$$

$$d^2 = \frac{k q_1 q_2}{F}$$

Now take the positive square root.

$$\sqrt{d^2} = \sqrt{\frac{k q_1 q_2}{F}}$$

$$d = \sqrt{\frac{k q_1 q_2}{F}}$$

We will not consider a problem of this type, but as always it is important and instructive to know how to perform the manipulation of the formulas we are discussing.

FRAME 10

Of greater importance than static electric charge is the movement of charge or current electricity.

If we could install a "window" in a copper wire and "watch" the charged electrons zip by with their 1.6019×10^{-19} coulomb of charge each, when 1 coulomb of charge was passing each second, we could define that quantity of moving charge as 1 ampere.

In other words, when 1 coulomb of charge or 6.24×10^{18} electrons pass the "window" in one second, one ampere of electricity is flowing.

FRAME 11

Now let's introduce the notion of a circuit. Better still, let's create one in our mind's eye. Suppose I have a conductor — a piece of copper wire will do. Bend it into an open loop. This is an open circuit. Now to get electrons to flow around the wire, we need something to give them a shove. In a "water circuit" (a pipe) we would use a pump or a high tower to give the potential.

In current electricity we will use a battery. The battery is a device that "pushes electrons" or has a difference of potential at each end, and the unit of measure is potential difference or the volt.

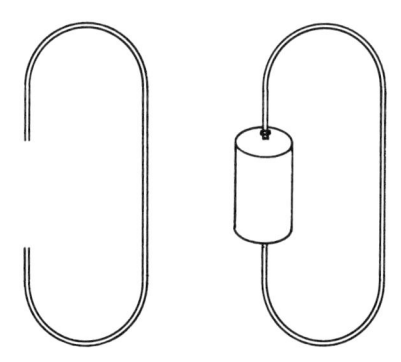

A *volt is defined as the work required to move a unit of charge from one place to another*, in other words, around the loop.

If work is measured in joules and charge in coulombs, then

$$1 \text{ volt} = \frac{1 \text{ joule}}{1 \text{ coulomb}}$$

When one joule of work is performed on 1 coulomb of charge to move it from one place to another, a potential difference of 1 volt exists between those two places.

FRAME 12

The evidence that it requires a potential difference or voltage to get the electrons to move around the wire gives us indirect evidence that there is resistance to the flow of electrons even in so-called conductors. This is true, and the unit of resistance is the *ohm*.

100 Magnetism, Electricity, and Power

The units of volt (v), ampere (i), and the ohm (r) have been stated in a nice concise law by a man named Ohm. This law states that *the current flowing in a circuit is directly proportional to the voltage applied and inversely proportional to the resistance.*

In symbols:

$$i = \frac{v}{r}$$

Where i is in amperes, v in volts, and r in ohms.

The quantities mentioned above are often associated with a schematic diagram that shows symbols for each.

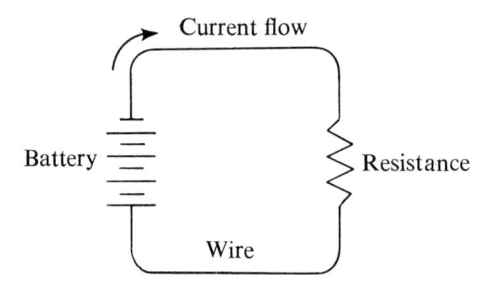

FRAME 13

91. The following schematic diagram is given. Find the unknown quantity.

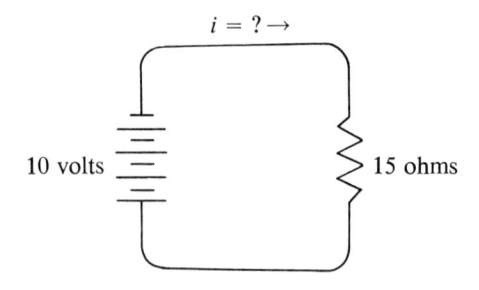

GIVEN	FIND
Volts: $v = 10$ volts	Current: $i = ?$
Ohms: $r = 15$ ohms	

Magnetism, Electricity, and Power   101

Our only formula to this point:

$$i = \frac{v}{r}$$

$$i = \frac{10 \text{ volts}}{15 \text{ ohms}}$$

$$i = \tfrac{2}{3} \text{ ampere}$$

FRAME 14

92. The following information is given. Draw a schematic diagram of a battery and resistor circuit and solve. Battery — 2.5 volts, resistance — 75 ohms. Find i.

FRAME 15

93. It is important that we can solve $i = v/r$ for v and for r. I will solve it for v and then you solve it for r.

GIVEN   FIND

$r \cdot i = \dfrac{v}{\cancel{r}} \cdot \cancel{r}$ solve for v: $v = ?$

$ir = v$

$v = ir$

FRAME 16

94. Solve Ohm's Law for r.

FRAME 17

95. A circuit has .01 amps flow with resistance of 1000 ohms. What is the voltage?

102 Magnetism, Electricity, and Power

GIVEN

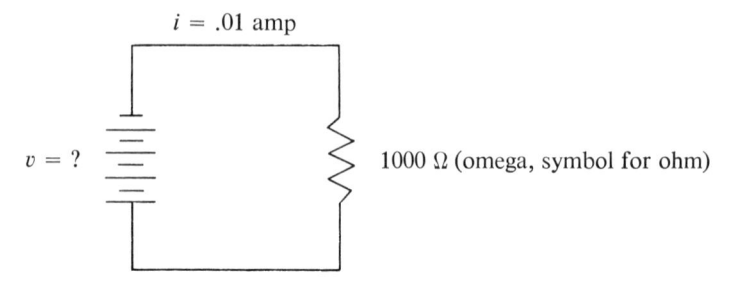

$i = .01$ amp

$v = ?$ 1000 Ω (omega, symbol for ohm)

FIND

Voltage: $v = ?$

$$v = ir$$
$$v = (.01)(1000)$$
$$v = 10 \text{ volts}$$

FRAME 18

96. What is the resistance in a circuit with a 6-volt battery and $\frac{1}{2}$-amp current flowing?

FRAME 19

There are two basic ways that resistance in a circuit can be arranged — in parallel and in series.

The following diagrams illustrate each one.

Parallel resistance

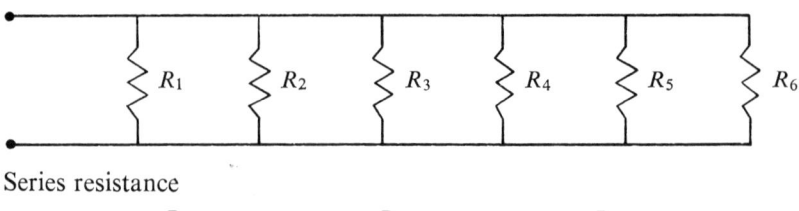

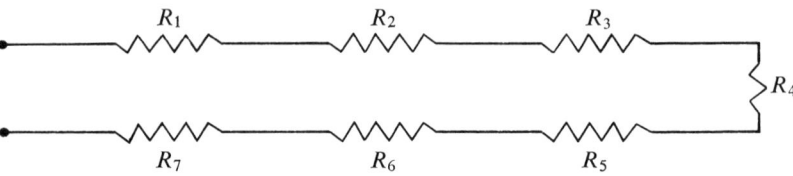

R_1 R_2 R_3 R_4 R_5 R_6

Series resistance

R_1 R_2 R_3

R_4

R_7 R_6 R_5

To add resistance in parallel, the following formula is necessary.

$$\frac{1}{R_t} = \frac{1}{R_1} + \frac{1}{R_2} + \frac{1}{R_3} + \frac{1}{R_4} + \frac{1}{R_5} \cdots$$

To add resistance in series, the following formula is necessary.

$$R_t = R_1 + R_2 + R_3 + R_4 + R_5 \cdots$$

FRAME 20

Consider a problem involving this formula.

97.

GIVEN

The following schematic diagram:

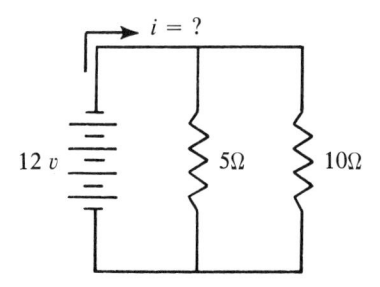

[Circuit diagram: 12 v battery connected to two parallel resistors of 5Ω and 10Ω, with $i = ?$ indicated]

FIND

Total current flowing: $i = ?$

The approach to the problem is to redraw the diagram with one equivalent resistor, then solve. Notice that the two resistors are in parallel so their sum is:

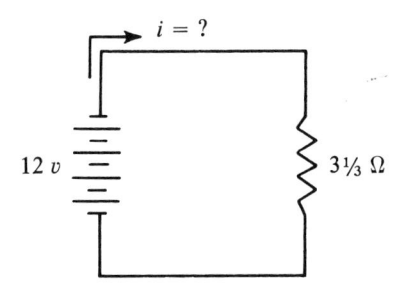

[Circuit diagram: 12 v battery connected to a single resistor of $3\frac{1}{3}\,\Omega$, with $i = ?$ indicated]

$$\frac{1}{R_t} = \frac{1}{R_1} + \frac{1}{R_2}$$

$$\frac{1}{R_t} = \frac{1}{5} + \frac{1}{10}$$

$$\frac{1}{R_t} = \frac{2}{10} + \frac{1}{10}$$

$$\frac{1}{R_t} = \frac{3}{10}$$

$$R_t = \frac{10}{3}$$

or

$$R_t = 3\frac{1}{3} \text{ ohms}$$

$$i = \frac{v}{r}$$

$$i = \frac{12}{3\frac{1}{3}}$$

$$i = 12 \div \frac{10}{3}$$

$$i = 12 \cdot \frac{3}{10} = \frac{36}{10} = 3.6 \text{ amps}$$

FRAME 21

Your problem:

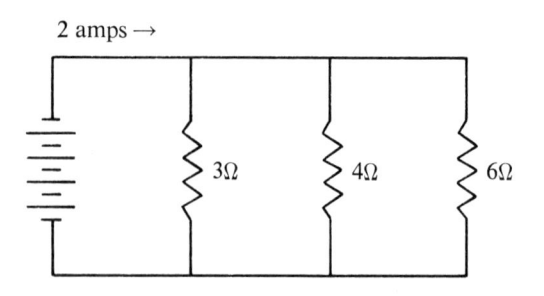

98.   GIVEN

2 amps →

[Circuit: battery with three parallel resistors 3Ω, 4Ω, 6Ω]

FIND

Voltage of battery: $v = ?$

FRAME 22

A problem involving series resistance is as follows:
99.   GIVEN

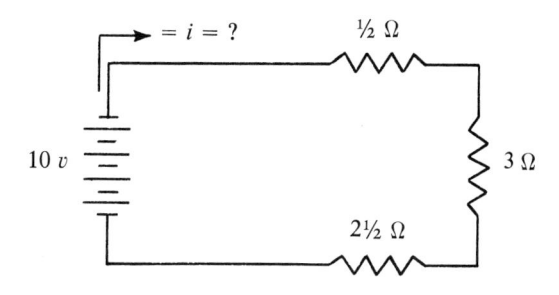

[Circuit diagram: 10 v battery connected in series with ½ Ω, 3 Ω, and 2½ Ω resistors; i = ?]

FIND

Current flow: $i = ?$

As in the other problems, we will resolve the problem to a single equivalent resistance and solve.

$$R_t = R_1 + R_2 + R_3$$
$$R_t = \frac{1}{2} + 3 + 2\frac{1}{2}$$
$$R_t = \frac{1}{2} + \frac{6}{2} + \frac{5}{2}$$
$$R_t = \frac{12}{2}$$
$$R_t = 6$$
$$i = \frac{v}{r}$$
$$i = \frac{10}{6}$$
$$i = \frac{5}{3} \text{ amp}$$

[Circuit diagram: 10 v battery connected to 6 Ω resistor; i = ?]

FRAME 23

100.   **GIVEN**

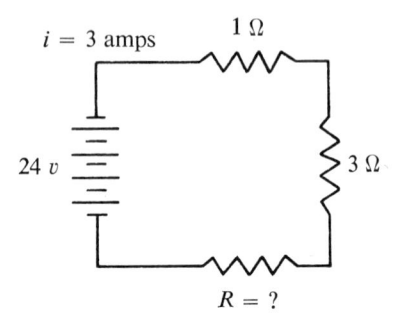

circuit diagram: 24 v battery, i = 3 amps, 1 Ω resistor, 3 Ω resistor, and R = ?

FIND

Resistance: $R_3 = ?$

FRAME 24

Earlier in this discussion we stated the definition of a volt. It was:

$$1 \text{ volt} = \frac{1 \text{ joule}}{1 \text{ coulomb}}$$

or, when 1 joule of work is performed on 1 coulomb of charge to move it from one place to another, a potential difference of 1 volt exists between the two places.

We can perform the following manipulation on the definition.

$$1 \text{ coulomb} \times 1 \text{ volt} = \frac{1 \text{ joule}}{1 \text{ coulomb}} \times 1 \text{ coulomb}$$

or

$$1 \text{ joule} = 1 \text{ coulomb} \times 1 \text{ volt}$$

Recall from our discussion of mechanical power that the unit of power was a joule per second or

$$\text{power (watt)} = \frac{\text{work (joule)}}{\text{time (second)}}$$

Make the substitution for work in this formula with joule = coulomb × volt.

The formula becomes

$$\text{watt} = \frac{\text{coulomb} \times \text{volt}}{\text{second}}$$

FRAME 25

Now recall another formula that we have discussed, the fact that an ampere is 1 coulomb of charge passing a point in one second.

$$\text{ampere} = \frac{\text{coulomb}}{\text{second}}$$

Look carefully at the formula

$$\text{watt} = \frac{\text{coulomb} \times \text{volt}}{\text{second}}$$

and replace the quantity coulomb/second with ampere.

$$\text{watt} = \text{ampere} \times \text{volt}$$

or in symbols

$p = iv$ the formula for electrical power.

Solve for i:

$$\frac{p}{v} = i\frac{\cancel{v}}{\cancel{v}}$$

$$i = \frac{p}{v}$$

Solve for v:

$$\frac{p}{i} = v\frac{\cancel{i}}{\cancel{i}}$$

$$v = \frac{p}{i}$$

FRAME 26

108 Magnetism, Electricity, and Power

101. Your stove uses 5 amps of electricity when it is on. It is serviced from a 220 volt line. How much power is being used?

GIVEN	FIND
Current: $i = 5$ amps	Power: $p = ?$
Voltage: $v = 220$ v	

$$p = vi$$
$$p = (220)(5)$$
$$p = 1100 \text{ watts}$$

FRAME 27

102. The power rating of your reading lamp is 100 watts. The voltage of the service is 110 v. Find the current flowing.

FRAME 28

Power consumption is measured in kilowatt hours. *A kilowatt hour is 1 kilowatt (1000 watts) being consumed for 1 hour or kilowatt × hours.* A practical problem in power consumption follows.

103. Suppose in your apartment you have the following electrical appliances. Toaster — 500 watts, hair dryer — 750 watts, iron — 300 watts, electric light — 100 watts. The operation each day of these appliances is as follows: Toaster — 15 minutes, hair dryer — $1\frac{1}{2}$ hours, iron — 1 hour, light — 4 hours. Determine the cost of operation at 4¢ per kilowatt hour.

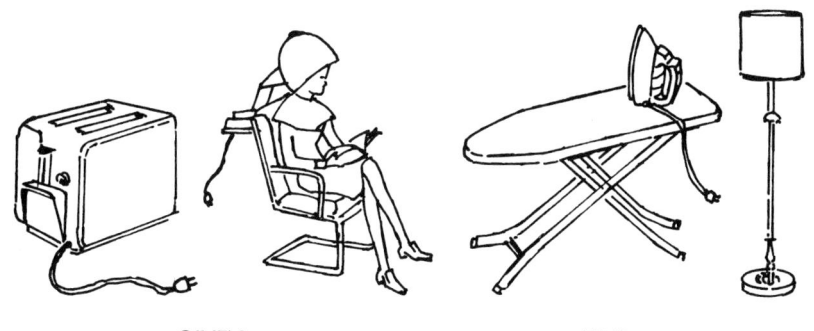

	GIVEN	FIND
Toaster	500 watts $\frac{1}{4}$ hour	Cost of operation
Hairdryer	750 watts $1\frac{1}{2}$ hours	
Iron	800 watts 1 hour	
Light	100 watts 4 hours	

1 kilowatt hour = 4¢

We need to change all the watts to kilowatts, then multiply by hours used.

$$.500 \text{ kwatts} \times .25 \text{ hours} = .125 \text{ kw hrs}$$
$$.750 \text{ kwatts} \times 1.50 \text{ hours} = 1.275 \text{ kw hrs}$$
$$.800 \text{ kwatts} \times 1.00 \text{ hours} = .800 \text{ kw hrs}$$
$$.100 \text{ kwatts} \times 4.00 \text{ hours} = .400 \text{ kw hrs}$$
$$\text{Total} \quad 2.400 \text{ kilowatt hours}$$

At 4¢ per kilowatt hour,

$$2.4 \times 4 = 9.6 \text{ cents.}$$

FRAME 29

Your problem:

104. You operate an electric water heater at 2500 watts for 6 hours, an electric clothes dryer at 1500 watts for 1 hour, and electric heat at 5000 watts for 8 hours. What is the cost at 4¢ per kilowatt hour?

9

Light and Relativity

FRAME 1

What makes the color red different from the color blue? Why can sunlight be divided into a spectrum of colors as it reflects from the beveled edge of a mirror? Is there a relationship between light and radio waves? These questions and many others are of immediate interest as we consider electromagnetic radiation.

The following diagram of the electromagnetic spectrum will be important to familiarize ourselves with relationships between different electromagnetic waves and the inherent properties of wavelength and frequency.

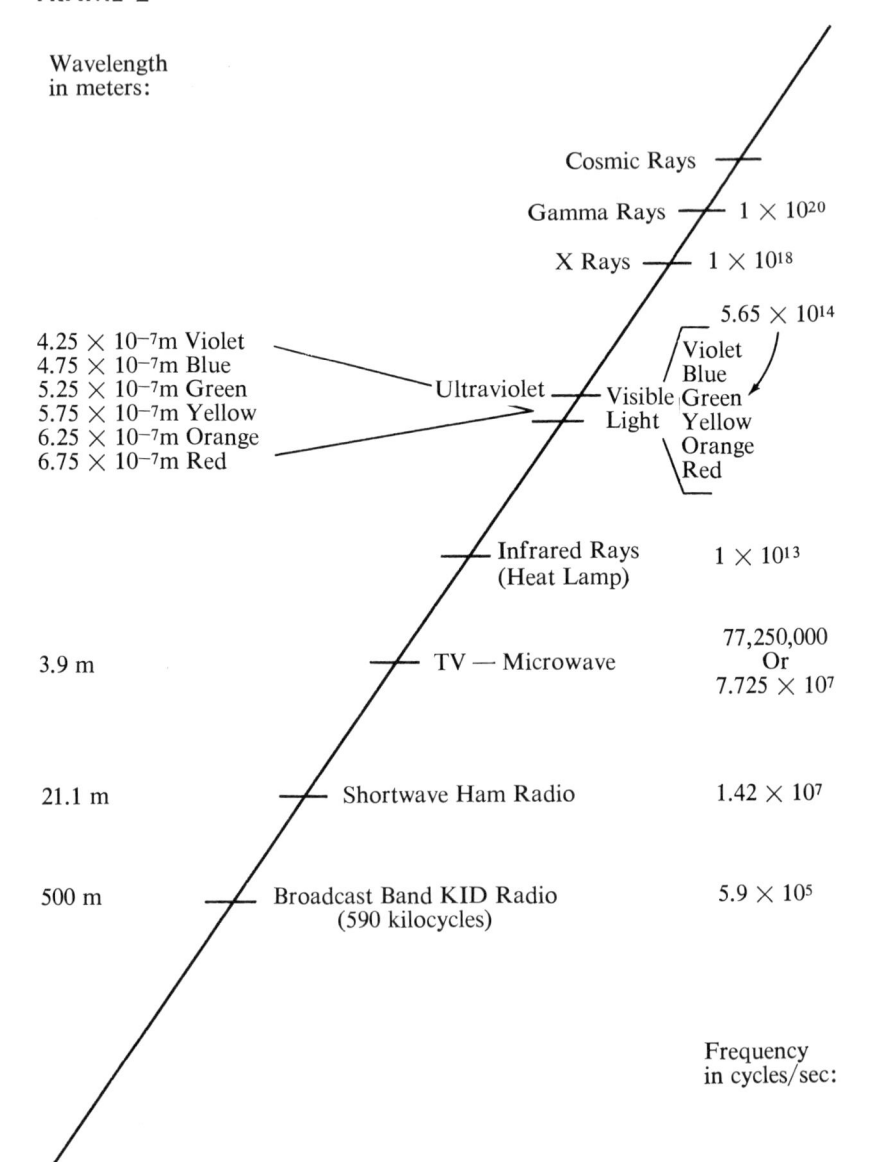

FRAME 2

Wavelength in meters:

                                              Cosmic Rays

                                    Gamma Rays 1×10^{20}

                                    X Rays 1×10^{18}

                                                        5.65×10^{14}

4.25×10^{-7}m Violet   Violet
4.75×10^{-7}m Blue   Blue
5.25×10^{-7}m Green   Ultraviolet   Visible Green
5.75×10^{-7}m Yellow   Light   Yellow
6.25×10^{-7}m Orange   Orange
6.75×10^{-7}m Red   Red

                                  Infrared Rays 1×10^{13}
                                  (Heat Lamp)

                                                                           77,250,000
3.9 m   TV — Microwave   Or
                                                                  7.725×10^7

21.1 m   Shortwave Ham Radio 1.42×10^7

500 m   Broadcast Band KID Radio 5.9×10^5
                                          (590 kilocycles)

                                                                   Frequency
                                                                   in cycles/sec:

FRAME 3

When a light is turned on in a dark room, the amount of light or illumination that will fall on a newspaper that we decide to pick up and read will depend on two things. First, the intensity of the light and second, the distance we are from the light.

The unit for measuring the illumination of our newspaper is the lumen and is measured in lumens per square foot or lumens per square meter. The intensity of the light is measured in candles (or candlepower). A 1 candle source is defined as the intensity of light emitted from a $\frac{1}{60}$ cm² hole to an interior that is maintained at 1755°C, the temperature of solidifying platinum. The amount of illumination incident on a surface 1 foot from a 1-candle source is 1 lumen per square foot.

The relationship between lumens, intensity in candles, and distance in feet is that the illumination in lumen/ft² is directly proportional to the intensity in candles and inversely proportional to the square feet. In symbols:

$$E \text{ (illumination)} = \frac{I \text{ (intensity)}}{R^2 \text{ (distance squared)}}$$

Solve for I:

$$R^2 \cdot E = \frac{I}{R^2} \cdot R^2$$

$$R^2 \cdot E = I$$

Solve for R:

$$R^2 \cdot E = \frac{I}{R^2} \cdot R^2$$

$$R^2 \cdot \frac{E}{E} = \frac{I}{E}$$

$$R^2 = \frac{I}{E}$$

$$R = \sqrt{\frac{I}{E}}$$

FRAME 4

105. What illumination, lumens/ft², will be provided on a newspaper 3 feet from a lamp with an intensity of 120 candles?

GIVEN	FIND
Distance: $R = 3$ ft	Illumination: $E = ?$

Intensity: $I = 120$ candles

Formula:

$$E = \frac{I}{R^2}$$

$$E = \frac{120}{(3)^2}$$

$$E = \frac{120}{9}$$

$$E = 13\tfrac{1}{3} \text{ lumens/ft}^2$$

FRAME 5

Your problem:

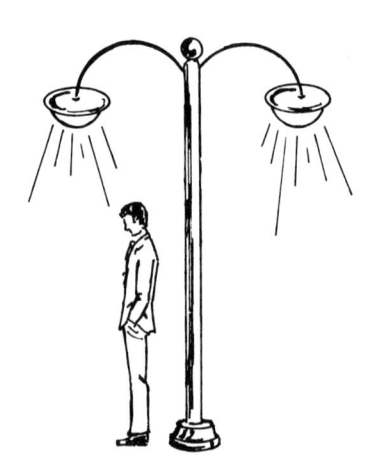

106. At 100 feet, what illumination will a street light of 2500 candles produce?

FRAME 6

107. Illumination of 50 lumens/ft² for reading a newspaper is recommended. How far should the paper be held from a 250-candle source?

               GIVEN   FIND
Illumination: $E = 50$ lumens/ft²   Distance: $R = ?$

Intensity: $I = 250$ candles

Recall from the formula the distance is:

$$R = \sqrt{\frac{I}{E}}$$

$$R = \sqrt{\frac{250}{50}}$$

$$R = \sqrt{5}$$

$$R = 2.2 \text{ feet}$$

FRAME 7

108. What is the intensity of a light, in candles, 2 meters away that illuminates .75 lumens/meter²?

FRAME 8

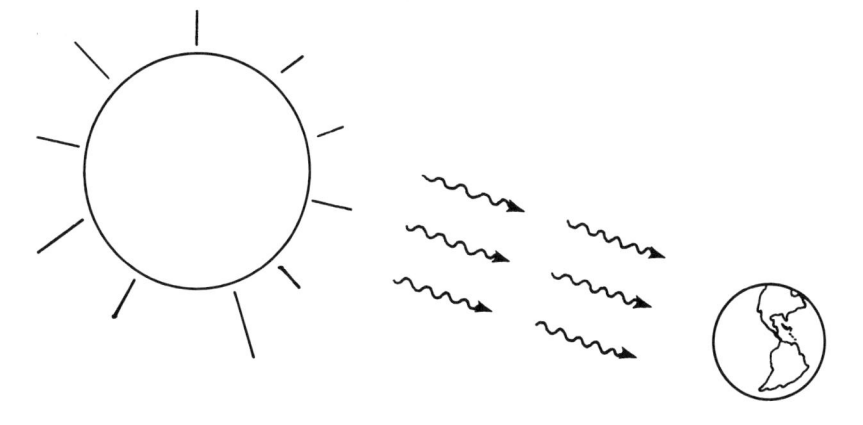

116 Light and Relativity

In our discussion of waves, we introduced the formula $v = \lambda \cdot f$. Recall that in our discussions of electromagnetic radiation, v was the velocity of a wave, λ the wavelength, and f the frequency.

We will need to keep the units consistent. If v is in ft/sec, then wavelength will need to be in ft/cycle and frequency in cycles/sec. Or if v is in m/sec, then wavelength is in m/cycle and frequency is in cycles/sec.

The v in the formula has been determined experimentally by Michelson and others as 3×10^8 m/sec or 186,000 miles/sec or 9.84×10^8 ft/sec.

The value of the speed of light is often abbreviated as c.

$$c = 3 \times 10^8 \, \frac{m}{sec}, \; 186,000 \, \frac{miles}{sec}, \text{ and } 9.84 \times 10^8 \, \frac{ft}{sec}$$

$$c = \lambda \cdot f$$

Solve for λ:

$$\frac{c}{f} = \frac{\lambda \cdot \cancel{f}}{\cancel{f}}$$

$$\frac{c}{f} = \lambda$$

$$\lambda = \frac{c}{f}$$

Solve for f:

$$\frac{c}{\lambda} = \frac{\cancel{\lambda} \cdot f}{\cancel{\lambda}}$$

$$\frac{c}{\lambda} = f$$

$$f = \frac{c}{\lambda}$$

FRAME 9

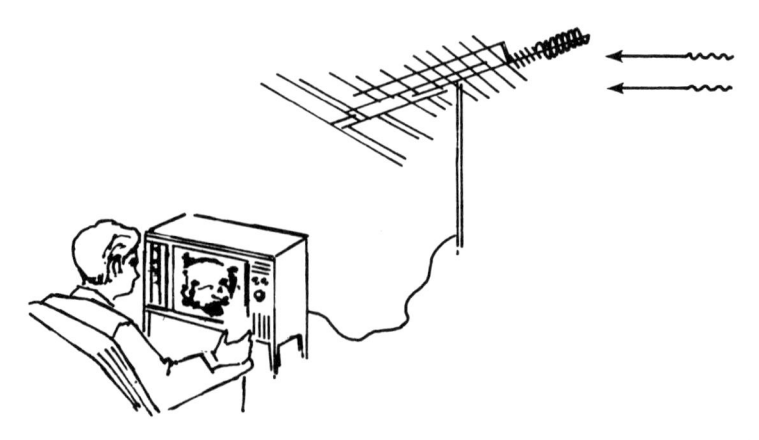

109. Channel 7 on your TV is broadcasting at 175.25 megacycles. What is the wavelength in meters?

Light and Relativity

GIVEN	FIND
Frequency:	Wavelength: $\lambda = ?$

$f = 175.25$ megacycle/sec

$f = 175.25 \times 10^6$ cycles/sec

$f = 1.75 \times 10^8$ cycles/sec
(rounded off)

Speed of light: $c = 3 \times 10^8$ m/sec

$$\lambda = \frac{c}{f} = \frac{3 \times 10^8 \text{ m/sec}}{1.75 \times 10^8 \text{ cycles/sec}} = \frac{3}{1.75} \times \frac{10^8}{10^8} = 1.7 \text{ meters/cycles}$$

FRAME 10

Your problem:

110. KSL radio is at 1160 kilocycles/sec on your radio dial. Find KSL wavelength in meters.

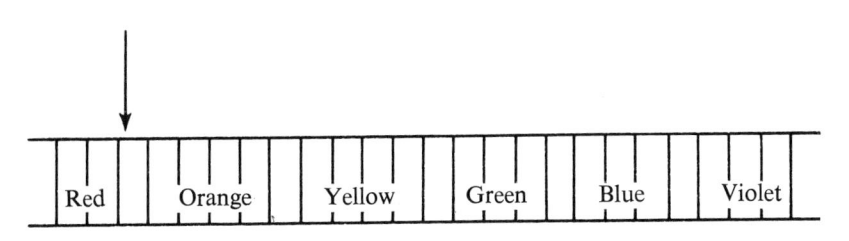

FRAME 11

| Red | Orange | Yellow | Green | Blue | Violet |

111. The wavelength of red light is 6.75×10^{-7} meters/cycles. Find the frequency in cycles/sec.

GIVEN	FIND
Speed of light: $c = 3 \times 10^8$ m/sec	Frequency: $f = ?$

Wavelength:

$\lambda = 6.75 \times 10^{-7}$ m/cycle

118 Light and Relativity

The formula: $f = c/\lambda$ Solve for f.

$$f = \frac{3 \times 10^8 \text{ m/sec}}{6.75 \times 10^{-7} \text{ m/cycle}} = \frac{3}{6.75} \times \frac{10^8}{10^{-7}} \text{ m/sec} \div \text{m/cycle}$$

$$= \text{m/sec} \times \text{cycle/m} \quad \text{(invert)}$$

$$f = .44 \times 10^{15} \text{ cycles/sec}$$

Answer: $f = 4.4 \times 10^{14}$ cycles/sec

FRAME 12

Your problem:

112. The wavelength of green light is 5.15×10^{-7} meters/cycle. Find the frequency.

FRAME 13

Early in the twentieth century Einstein, Lorentz and others worked on the ideas of relativity and formulated a general and a special theory. We will consider some ideas from the theory of special relativity. The theory has some verification with high speed particles. In this case, high speed refers to a significant fraction of the speed of light.

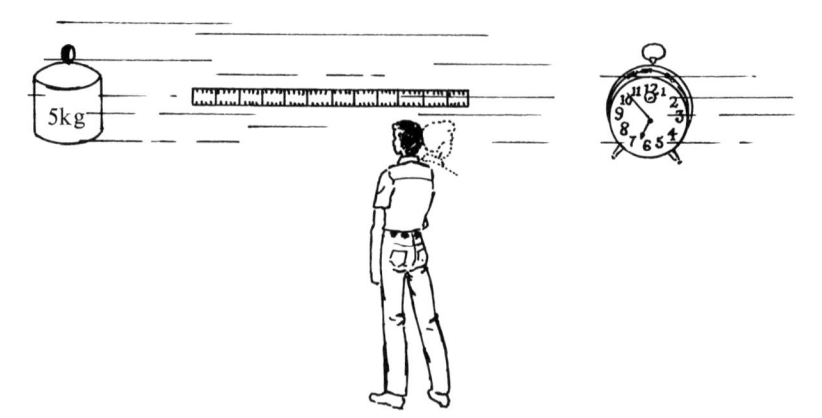

The following quantities are hypothesized to be altered when moving at high speeds with respect to a stationary observer. Perhaps at this point it could be stated that a passenger riding in some fantastic rocket moving at these tremendous speeds is not aware of any alterations. Only these quantities are altered: time is dilated or slowed down, length is shortened, mass is increased. The following formulas are the summary of the special theory of relativity.

Time: $t = \dfrac{t_o}{\sqrt{1 - \dfrac{v^2}{c^2}}}$

Length: $l = l_o \sqrt{1 - \dfrac{v^2}{c^2}}$

Mass: $m = \dfrac{m_o}{\sqrt{1 - \dfrac{v^2}{c^2}}}$

Where t_o, l_o, m_o are measures of the objects at rest, and t, l, m are the measures at high velocity.

v = velocity of the moving quantity
c = velocity of light

FRAME 14

A careful observation of each formula will reveal the quantity:

$$\sqrt{1 - \dfrac{v^2}{c^2}}$$

which occurs in each formula. We will consider this quantity in detail to see how different values of v affect each problem.

To start, let's call $\sqrt{1 - (v^2/c^2)} = B$ and replace it in each formula

Time: $t = \dfrac{t_o}{B}$

Length: $l = l_o \cdot B$

Mass: $m = \dfrac{m_o}{B}$

Now let's suppose that $v = 0$ or we have a situation in which the object is at rest. Let's consider what effect the formula has on time, length, and mass.

Evaluate $\sqrt{1 - (v^2/c^2)}$ when $v = 0$ m/sec

$$c = 3 \times 10^8 \text{ m/sec}$$

Substitute

$$\sqrt{1 - \frac{(0 \text{ m/sec})^2}{(3 \times 10^8 \text{ m/sec})^2}}$$

$$\sqrt{1 - 0}$$

$$\sqrt{1}$$

$$1$$

Therefore: $B = 1$ when $v = 0$ m/sec

Replace $B = 1$ in each formula.

Time: $t = \dfrac{t_o}{B}$ $t = \dfrac{t_o}{1}$ $t = t_o$

Length: $l = l_o \cdot B$ $l = l_o \cdot 1$ $l = l_o$

Mass: $m = \dfrac{m_o}{B}$ $m = \dfrac{m_o}{1}$ $m = m_o$

In conclusion, we can state that when a body is not moving, its rest mass, length, and time are as we observe them (just as we would expect).

FRAME 15

Let's now consider the other extreme.

Suppose $v = 3 \times 10^8$ m/sec, an impossibility at present but an interesting idea.

Now substitute. $v = 3 \times 10^8$ m/sec

$c = 3 \times 10^8$ m/sec in the formula B

$$B = \sqrt{1 - \frac{v^2}{c^2}} = \sqrt{1 - \frac{(3 \times 10^8 \text{ m/sec})^2}{(3 \times 10^8 \text{ m/sec})^2}} = \sqrt{1 - 1} = \sqrt{0} = 0$$

Now substitute $B = 0$ into each of the formulas. Recall $B = 0$ because $v = 3 \times 10^8$ m/sec.

Time: $t = \dfrac{t_o}{B}$ $t = \dfrac{t_o}{0}$ $t = ?$

Length: $l = l_o \cdot B$ $l = l_o \cdot 0$ $l = 0$

Mass: $m = \dfrac{m_o}{B}$ $m = \dfrac{m_o}{0}$ $m = ?$

Light and Relativity 121

What is the meaning of t and m? Suppose we look at this series of fractions.

$$\frac{12}{12}, \frac{12}{6}, \frac{12}{4}, \frac{12}{3}, \frac{12}{2}, \frac{12}{1}, \frac{12}{\frac{1}{2}}, \frac{12}{\frac{1}{3}}, \ldots \frac{12}{0}$$

$$\frac{12}{12} = 1, \frac{12}{6} = 2, \frac{12}{4} = 3, \frac{12}{3} = 4, \frac{12}{2} = 6, \frac{12}{1} = 12, \frac{12}{\frac{1}{2}} = 24, \frac{12}{\frac{1}{3}} = 36, \ldots \infty$$

As the numerator remains 12 and the denominator gets smaller, what happens to the value of the fraction? The answers to the fractions 1, 2, 3, 4, 6, 12, 24, 36, ... approach infinity, the value for $\frac{12}{0}$.

Conclusion: Therefore v approaches ∞, m approaches ∞ and $l = 0$ when the velocity equals the speed of light. What is the meaning of this? The Lorentz transformations indicate that as a body travels at the speed of light, each second of time drags out to be infinitely long until time stops, length shortens to 0, and mass becomes so large as to be infinite. These ideas become philosophical at this point don't they?

FRAME 16

Now that we have considered both extremes, let's look at a speed in the realm of possibility.

113. Suppose you were traveling 2×10^8 m/sec and had a 5 kg mass and a clock that elapsed 2 seconds. How would these quantities appear to the stationary observer as you whizzed by?

GIVEN	FIND
$m_o = 5$ kg	$t = ?$
$t_o = 2$ seconds	$m = ?$
$v = 2 \times 10^8$ m/sec
$c = 3 \times 10^8$ m/sec

$$t = \frac{t_o}{\sqrt{1 - \frac{v^2}{c^2}}} \quad \text{and} \quad m = \frac{m_o}{\sqrt{1 - \frac{v^2}{c^2}}}$$

122 Light and Relativity

Let's evaluate the common expression $\sqrt{1 - (v^2/c^2)}$

$$\sqrt{1 - \frac{v^2}{c^2}} = \sqrt{1 - \frac{(2 \times 10^8)^2}{(3 \times 10^8)^2}} = \sqrt{1 - \frac{4 \times 10^{16}}{9 \times 10^{16}}} = \sqrt{1 - \frac{4}{9}}$$

$$= \sqrt{\frac{9}{9} - \frac{4}{9}} = \sqrt{\frac{5}{9}} = \frac{\sqrt{5}}{3} = \frac{2.23}{3} = .74$$

Now let's solve each equation using $.74 = \sqrt{1 - (v^2/c^2)}$

$$t = \frac{2 \text{ seconds}}{.74} = 2.7 \text{ seconds}$$

$$m = \frac{5 \text{ kg}}{.74} = 6.8 \text{ kg}$$

The seconds appear to be dragged out, thus slowing time, and the mass appears to increase.

FRAME 17

Your problem:

114. Suppose you are traveling 2×10^8 m/sec and have a 10-meter stick pointing in the direction of travel. How long does it appear to the stationary observer?

10

Cathode Rays, X-Rays, Radioactivity

FRAME 1

One approach to understanding particle physics is to trace the development of our understanding of the electron. Early experiments by Crookes, Millikan, and Thomson announced the discovery of a particle of mass, $m = 9.1 \times 10^{-31}$ kg and a negative charge, $e = -1.6 \times 10^{-19}$ coulomb. It was named the electron. Thomson's experiments with electrons in cathode ray tubes pointed out that as the voltage of the tube was increased, the energy of accelerated electrons could also be increased.

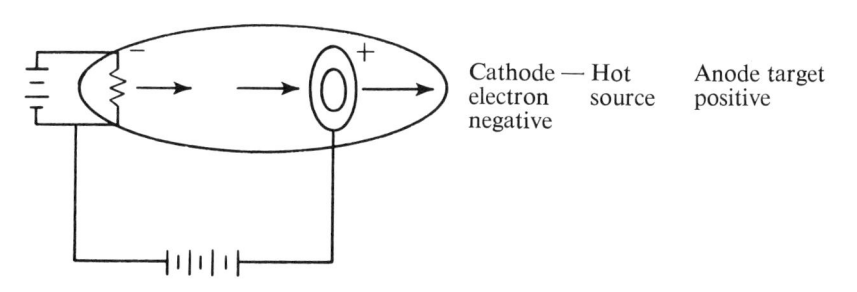

Cathode — Hot electron source negative
Anode target positive

Kinetic energy of the electrons was equal to voltage applied between cathode and anode and the charge on the electron.

In symbols:

$$\tfrac{1}{2}mv^2 = eV \quad \text{or} \quad \frac{mv^2}{2} = eV$$

Here $\frac{1}{2}mv^2$ is the Kinetic energy (m = mass, v = velocity) and eV is equivalent in units to e (charge) V (volt). Thus the unit of electron volts is equivalent to joules of energy.

FRAME 2

Let's solve $mv^2/2 = eV$ for V and v.

Solved for V:

$$\left(\frac{1}{e}\right)\left(\frac{mv^2}{2}\right) = (eV)\left(\frac{1}{e}\right)$$

$$\frac{mv^2}{e2} = \frac{eV}{e}$$

$$\frac{mv^2}{2e} = V$$

Solved for v:

$$2\left(\frac{mv^2}{2}\right) = (eV)(2)$$

$$\frac{mv^2}{m} = \frac{eV2}{m}$$

$$v^2 = \frac{2eV}{m}$$

$$v = \sqrt{\frac{2eV}{m}}$$

FRAME 3

115. The electrons in a TV picture tube are traveling $.5 \times 10^8$ m/sec ($\frac{1}{16}$ the speed of light). What voltage is being applied?

GIVEN

Velocity: $v = .5 \times 10^8$ m/sec

Mass: $m = 9.1 \times 10^{-31}$ kg

Charge: $e = 1.6 \times 10^{-19}$ coulomb

FIND

Voltage: $V = ?$

As we look at the three formulas, the one solved for the unknown is:

$$V = \frac{mv^2}{2e}$$

$$V = \frac{(9.1 \times 10^{-31} \text{ kg})(.5 \times 10^8 \text{ m/sec})^2}{(2)(1.6 \times 10^{-19} \text{ coulomb})}$$

$$V = \frac{(9.1)(.25)}{(2)(1.6)} \times \frac{(10^{-31})(10^{16})}{10^{-19}} = \frac{2.28}{3.2} \times 10^4 = .7 \times 10^4 = 7. \times 10^3$$

$$V = 7 \times 10^3$$

FRAME 4

Your problem:

116. The voltage that gives energy to an electron beam is 2500 volts. What is the velocity of the electrons?

FRAME 5

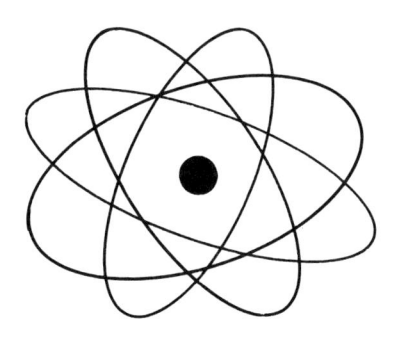

Another field of electron experiment is with the atom. Early in the 20th century, a famous Danish scientist, Niels Bohr, gave the world a mechanical theory of the atom. He visualized each atom as a tiny planetary system with a heavy nucleus at the center (the sun) and electrons (the planets) revolving about the nucleus.

Experiments on atoms performed by Bohr and others revealed that electrons could be removed from an inner orbit of low energy and forced to an outer orbit of high energy by means such as an electrical spark. The spark knocked an electron from its ground state orbit to an excited orbit by giving it energy. In the new orbit, the atom is unstable and the electron will cascade back to the low energy vacancy and in the process, give up the gained energy. The energy that these excited atoms emit is due to their orbital electrons returning to normal or ground state and is in the form of light.

In symbols:

$$E_1 - E_2 = h\nu$$

where E_1 is the high energy orbit (outer) in joules, and E_2 is the low energy orbit (inner) in joules. The difference between these two energies is emitted as light,

where h = Planck's constant (proportionality constant)

ν = Frequency of the emitted light

FRAME 6

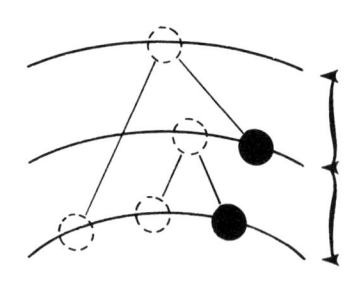

117. The difference in energy in joules between the excited and ground states of an electron is 3.74×10^{-19} joules. What frequency of light will it emit when the excited electron returns to ground state?

GIVEN · FIND

Difference:

$E_1 - E_2 = 3.74 \times 10^{-19}$ joules Frequency: $\nu = ?$

Planck's constant:

$h = 6.62 \times 10^{-34}$ joules·sec

Cathode Rays, X-Rays, Radioactivity 127

$$\frac{E_1 - E_2}{h} = \frac{h\nu}{h}$$

$$\frac{E_1 - E_2}{h} = \nu$$

$$\nu = \frac{3.74 \times 10^{-19}}{6.62 \times 10^{-34}} = .565 \times 10^{15} = 5.65 \times 10^{14} \text{ cycles/sec}$$

FRAME 7

Your problem:

118. A *photon* of frequency 4.38×10^{14} cycles/sec (red light) is emitted as an electron cascades back to ground state. What is the difference in the ground state and excited state energy levels in joules?

FRAME 8

To this point we have discussed the energy that a voltage difference can give to an electron in a cathode ray tube by the formula $\frac{1}{2}mv^2 = Ve$. Our discussion has also considered the energy difference manifested in the form of light or electromagnetic radiation as an electron jumps between two orbits or energy levels in the atom. That equation is $E_1 - E_2 = h\nu$. An obvious question at this point is, Could an equation be formed $h\nu = Ve$ concerning the energies of electrons? The answer is, Yes, and the physical phenomena is called X-ray production.

Two phenomena have been observed in X-ray production. The first occurs when a high energy electron approaches an atom, penetrates it, knocks out an inner orbit electron and other electrons cascade to the lower energy level, giving off energy in the process. The emitted energy is extremely high-frequency electromagnetic radiation called X-ray.

FRAME 9

The second way that an atom can emit an X ray is when a high velocity electron approaches an atom, penetrates it and does not collide with any other electrons but is deflected by the strong positive charge of the nucleus. During the deflection, a wave of EMR of energy $h\nu$ is emitted. Since the Law of Conservation of Momentum must hold in this type of "collision," the electron is deflected to one side and the $h\nu$ energy wave of EMR is emitted from the other side.

128 Cathode Rays, X-Rays, Radioactivity

The acceleration of electrons in an X-ray tube and the maximum frequency of the emitted wave is given by the following formula.

$$h\nu_{max} = Ve$$

This is the formula we assumed was possible in this discussion.

$h =$ Planck's constant
$\nu =$ frequency of emitted wave
$V =$ voltage of X-ray tube
$e =$ charge of electron

Solve $\dfrac{h\nu_{max}}{e} = \dfrac{V\cancel{e}}{\cancel{e}}$ for V. Solve $\dfrac{\cancel{h}\nu_{max}}{\cancel{h}} = \dfrac{Ve}{h}$ for ν.

$\dfrac{h\nu_{max}}{e} = V$ $\nu_{max} = \dfrac{Ve}{h}$

FRAME 10

119. The voltage across an X-ray tube is 30,000 volts. What is the maximum possible frequency of an X ray produced in the tube?

GIVEN   FIND
Voltage: $V = 30{,}000$ volts   Frequency of X-ray: $\nu_{max} = ?$
Planck's constant: $h = 6.63 \times 10^{-34}$
Charge on electron:
$e = 1.6 \times 10^{-19}$ coulomb

$$\nu_{max} = \dfrac{Ve}{h}$$

$$\nu_{max} = \dfrac{(3. \times 10^4)(1.6 \times 10^{-19})}{6.63 \times 10^{-34}}$$

$$\nu_{max} = \dfrac{(3)(1.6)}{6.63} \times \dfrac{(10^4)(10^{-19})}{10^{-34}}$$

$$\nu_{max} = \dfrac{4.8}{6.63} \times 10^{19}$$

$$\nu_{max} = .725 \times 10^{19}$$

$$\nu_{max} = 7.25 \times 10^{18} \text{ cycles/sec}$$

FRAME 11

Your problem:

120. What voltage exists across an X-ray tube when X rays of 5×10^{18} cycles/sec are emitted?

FRAME 12

Our discussions to this point can be concluded in the following equation:

$$E = h\nu = \tfrac{1}{2}mv^2 = Ve$$

An electron can manifest its energy as light, kinetic energy, or as the product of voltage times charge as it is accelerated across an X-ray tube.

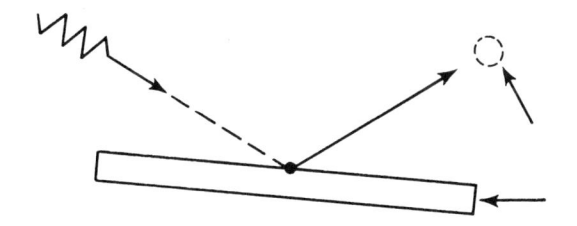

X-ray production involves a high velocity electron striking an atom and the resultant electromagnetic wave emitted. Of great interest and of significant importance is the reverse phenomena. As an electromagnetic wave strikes an atom (plate of alkali metals), an electron is emitted. This phenomena has given great strength to the particle theory of light or the photon theory. Light can be thought of as discrete packets of energy and will be "absorbed" by an electron thus giving it extra energy and dislodging it.

Einstein formulated the photoelectric effect in this way:

$$h\nu = w + \tfrac{1}{2}mv^2$$

Or in other words:

energy of incident photon = work to "dig" out electron
+ kinetic energy of the dislodged electron.

FRAME 13

121. Suppose an incident photon on a metal has energy just equal to the work function of the metal. What will be the kinetic energy of the dislodged electron?

GIVEN FIND
Equal photon and work function: Kinetic energy: $\frac{1}{2}mv^2 = ?$

$$hv = w$$

Formula:

$$hv = w + \tfrac{1}{2}mv^2$$

If the quantities hv and w are equal, then $\frac{1}{2}mv^2$ will have to equal 0 to maintain equality.

FRAME 14

122. A metal has a work function that is very small. Suppose $w = 0$ for practical purposes. What does the photoelectric formula then look like?

FRAME 15

Our discussion of the electron continues as we look at radioactivity.

In the late 1800's, it was discovered that certain naturally occurring elements spontaneously emit some type of radiation. This radiation was observed to expose photographic film, ionize gases and cause some other physical phenomena. For lack of a better word, this radiation was called radioactivity. It was noted that the elements that emit radiation could not be changed *in any way* to disturb this constant discharge of particles.

After many experiments, it was discovered that three basic types of radiation occur. These can be distinguished best in an electric field. The following diagram helps explain this observation.

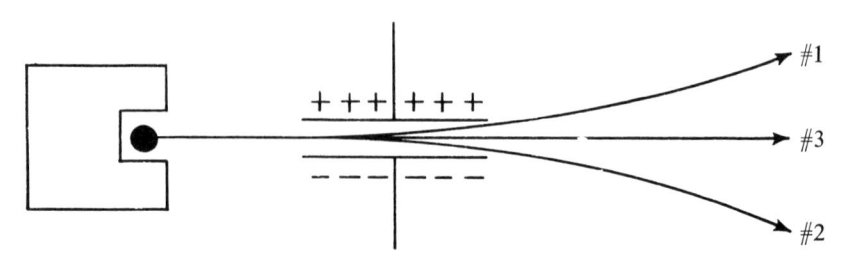

Those particles that are attracted toward the +, positive plate, #1, are negatively charged (unlike charges attract) and are called beta rays or just our old friend, an emitted electron. Those attracted toward the −, negative plate, #2, are positively charged and called alpha rays. Those that are not deflected, #3, are called gamma rays and have no charge.

FRAME 16

The following information has been discovered about each of these particles:

	Description	Velocity	Mass	Symbol
Alpha:	2 protons and 2 neutrons emitted as a single particle. A helium nucleus.	10,000 m/sec (slow)	7,000 times heavier than an electron	$_{+2}He^4$ or $_{+2}\alpha^4$
Beta:	electron	$\frac{1}{2}$ c.	9.1×10^{-31} kg	$_{-1}e^0$ or $_{-1}\beta^0$
Gamma:	EMR — very high frequency. Higher than X-rays.	speed of light	none (just energy)	γ

FRAME 17

An explanation of where these particles originate is now needed. Some atoms kick out of their nucleus fragments of their own nucleus. These are the radioactive atoms. These fragments that are spontaneously emitted are what we detect as alpha and beta particles. The gamma ray that we detect is simply emitted energy in the form of electromagnetic radiation. When alphas, betas and gammas are emitted, the original atom becomes a different atom because of the loss of part of its nucleus. Associated with each of these radioactive elements is a length of time called "half-life." This is the time necessary for half of any quantity of atoms to change to different atoms by emission of radioactive particles.

FRAME 18

Consider the following problem:

123. Thorium 234. The number 234 is the mass number of Thorium and is the total sum of all neutrons and protons. The symbol for Thorium 234 is:

$$Th^{234}$$

Let's suppose we have 96 atoms of this element and the half-life is 25 days as determined experimentally. After 25 days, 48 of the original 96 atoms ($\frac{1}{2}$ of them) will have emitted a beta particle and changed or decayed to Protactinium leaving 48 Thorium atoms. After 25 more days have passed, of the remaining Thorium atoms, 24 of them will have decayed to Protactinium leaving 24 Thorium. This continues. Diagrammatically:

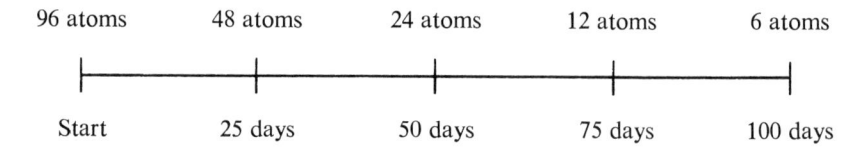

96 atoms	48 atoms	24 atoms	12 atoms	6 atoms
Start	25 days	50 days	75 days	100 days

FRAME 19

Consider another problem:

124. Suppose you have 64 grams of Bi^{210} (Bismuth 210) with a half-life of 5 days. How many atoms of Bismuth will you have in 20 days?

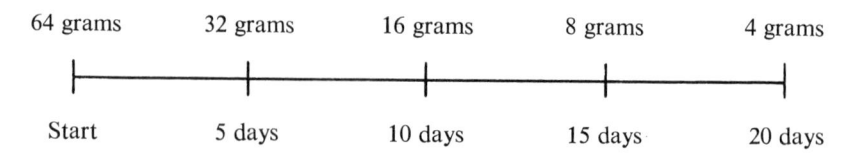

64 grams	32 grams	16 grams	8 grams	4 grams
Start	5 days	10 days	15 days	20 days

FRAME 20

Work these 2 problems.

125. If 1 kg of radium is sealed into a container, how much of it will remain as radium after (a) 1600 years (b) after 4800 years? (The half-life of radium is 1600 years.)

FRAME 21

126. You order a radioactive isotope from a supplier. When the isotope arrives through the mail 4 days later (96 hrs), there are only 2 grams of your desired isotope remaining. The half-life is 24 hours. How much did the supplier originally mail?

FRAME 22

In order to understand the nature of the atom by using radioactive emissions, consider the following experiment. Ernest Rutherford, an experimental physicist, about the turn of the century performed the following experiment:

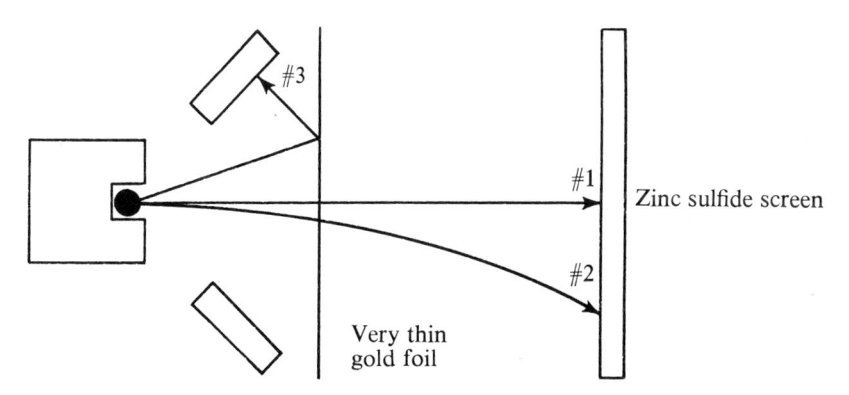

Conclusions:

1. Atoms are mostly empty space because most alpha particles pass through the foil with no deflection (position 1 on the screen).

2. Occasionally an alpha particle is deflected indicating it came near to a particle in the foil (position 2 on the screen).

3. On rarer occasions one of the alpha particles reflects or hits something very solid and bounces back (position 3 on the screen).

From Rutherford's work emerges our modern concept of an atom: a small, heavy core or nucleus, containing all the atom's positive charge and most of its mass, surrounded by the electron cloud, a group of electrons at relatively great distances from the nucleus and one another.

Experimentation has provided the general information that the nucleus is composed of protons and neutrons with charges positive and neutral re-

134 Cathode Rays, X-Rays, Radioactivity

spectively. Further, we denote the number of protons in the nucleus of an atom as its atomic number.

FRAME 23

With this concept, we can now consider what happens in the radioactive decay of particular atoms. There are several types:

1. alpha emission
2. beta emission (2 types)
3. gamma emission

Let's consider the following example of alpha decay:

$$_{88}Ra^{226} \rightarrow {}_2\alpha^4 + {}_{86}Rn^{222}$$

Consider this example of alpha decay:

$$_{92}U^{238} \rightarrow {}_2He^4 + {}_?\boxed{}^?$$

Step 1: Step 2: Step 3:

$_{90}\boxed{} \xrightarrow{then} {}_{90}\boxed{}\,234 \xrightarrow{then} {}_{90}\boxed{Th}\,234$

FRAME 24

Now let's consider a beta decay:

$$_{83}Bi^{210} \rightarrow {}_?\boxed{}^? + {}_{84}Po^{210}$$

$$_?\boxed{}^? = {}_{-1}\boxed{e}^0 \quad \text{Electron}$$

Another type of beta decay is:

$$_{13}Al^{26} \rightarrow {}_{+1}e^0 + {}_?\boxed{}^?$$

$$_?\boxed{}^? = {}_{12}\boxed{Mg}\,26 \quad \text{Magnesium}$$

Note: Positron emission

Cathode Rays, X-Rays, Radioactivity

FRAME 25

Solve these equations by determining the unknown quantity.

127. $_{90}Th^{232} \rightarrow {_2}He^4 + \boxed{?}$
128. $_{83}Bi^{210} \rightarrow {_{-1}}e^0 + \boxed{?}$
129. $_{89}Ac^{228} \rightarrow \boxed{?} + {_{87}}Fr^{224}$
130. $_{20}Ca^{45} \rightarrow \boxed{?} + {_{21}}Sc^{45}$

FRAME 26

Similar reactions to those we just considered involve the bombarding of an element by a particle and transforming it into another element. This process is called transmutation. An example follows:

$$_{5}B^{11} + {_2}He^4 \rightarrow {_0}n^1 + {_7}N^{14}$$

$$_{4}Be^9 + {_1}H^1 \rightarrow {_1}H^2 + {_4}Be^8$$

FRAME 27

Solve the following problems by determining the missing quantity.

131. $_{11}Na^{23} + {_2}He^4 \rightarrow {_1}H^1 + ?$
132. $_{14}Si^{28} + {_0}n^1 \rightarrow {_1}H^1 + ?$
133. $_{5}B^{11} + {_1}H^1 \rightarrow {_2}He^4 + ?$

Answers to Student's Problems

Unit 1
2. 50 miles/hour
4. 1,000 miles
6. $1\frac{1}{2}$ hours
8. 5 miles/hour·sec
10. -120 ft/sec^2
12. 39.2 meters/sec
14. 3 seconds
16. for $T = 2$ sec, $v = 64$ ft/sec, $d = 64$ ft
    for $T = 3$ sec, $v = 96$ ft/sec, $d = 144$ ft
    for $T = 4$ sec, $v = 128$ ft/sec, $d = 256$ ft
    for $T = 4.5$ sec, $v = 144$ ft/sec, $d = 324$ ft

Unit 2
18. $\frac{1}{3}$ meters/sec^2
21. mass varies according to individual student's weight
23. 78.4 newtons
25. -3500 lbs
27. 75 lbs

Unit 3
29. (18,9)
31. 5 lbs
32. one component is parallel to the ground pointing downhill and one component is perpendicular to the ground, pointing into the hill.
34. $3\frac{3}{4}$ lbs at A; $1\frac{1}{4}$ lbs at B

Unit 4
36. 25 lbs
38. 2 kgs
40. the gravitational force increased from $\frac{1}{4}$ to 1, an increase of 4 times.
42. the force increased from $\frac{1}{2}$ in Situation 1 to 4 in Situation 2, an increase of 8 times.

Unit 5
44. 97,500 ergs
46. $6\frac{2}{3}$ lbs
48. $\frac{3}{4}$ hp
50. 2 hp
52. $\frac{1}{2}$ hp
54. 80 ft·lbs
56. 50 ft·lbs
58. 4 slugs

Unit 6
60. $4\frac{1}{2}$ lbs/in^2
62. a total of 200 in^2 of rubber is in contact with the road

Answers to Student's Problems   137

64. 10 lbs/ft³
66. 6,200,000 lbs of grain
68. 45/62 of its height
70. 261.8 cycles/sec
72. 3.3 × 10⁻⁷ meters/cycle

Unit 7

74. −44 4/9°
76. 258° K
78. 360 kilo/cal of heat
80. 10 ft³
82. 546° K
84. 75° K
86. ⅔ atmosphere

Unit 8

88. The force in Situation 2 is twice as great as in Situation 1.
90. 7 × 10⁻³ newtons
92. .03 amps
94. $r = \dfrac{v}{i}$
96. 12 ohms
98. 2⅔ volts
100. 4 ohms
102. .91 amps
104. $2.26

Unit 9

106. .25 lumens/ft²
108. 3 candles
110. 258 meters
112. 5.8 × 10¹⁴ cycles/sec
114. 7.4 meters

Unit 10

116. 2.96 × 10⁷ meters/sec
118. 29 × 10⁻²⁰ joules
120. 2.07 × 10⁴ volts
122. $hv = \tfrac{1}{2} mv^2$
125. for 1600 years — ½ kilogram
     for 4800 years — ⅛ kilogram
126. 32 grams
127. $_{88}Ra^{228}$
128. $_{84}Po^{210}$
129. an alpha particle
130. an electron
131. $_{12}Mg^{26}$
132. $_{13}Al^{28}$
133. $_{4}Be^{8}$

Index

Acceleration, 6, 18, 25
  definition, 7
  negative, 9
  units, 8
Alpha rays, 131, 133
Ampere, 98
Amplitude, 74
Archimedes' Principle, 71
Area, 66
Atoms, 93, 133
Attraction, 94

Balloons, 90
Battery, 99, 104
Beta decay, 134
Beta rays, 131
Bohr, 125
BTU, 82
Buoyancy, 71

Calories, 82
Candle power, 113
Cathode rays, 123
Centripetal force, 39, 40
Charge
  positive, 93
  negative, 93
  neutral, 93
Charles' Law, 86, 89
Circular motion, 39
Color, 111, 117
Conductor, 98
Conservation of momentum, 127
Constant of proportionality, 95
Coulombs' Law, 95
Crookes, 123
Current, 98

Deceleration, 9
Density, 68
  table of common substances, 69
Directly proportional, 87, 94
Displaced fluid, 72
Displacement, 50
Distance, 4
  units, 4

Einstein, 118, 129
Electric current, 98

Electron, 33, 93, 133
  charge, 123
  mass, 123
Electromagnetic spectrum, 111
Electron cloud, 133
Electrostatics, 97
Energy
  chemical, 57
  kinetic, 57, 59
  potential, 57
Equations, 5
  solving for unknown, 5
  list of —
    $a = f/m$, 18
    $a = V_2 - V_1/t$, 7
    area = length·width·height, 66, 68
    buoyant force = weight of displaced fluid, 71
    $c = \lambda f$, 116
    density = weight/volume, 69
    $d = \frac{1}{2} gt^2$, 15
    $E = I/R^2$, 113
    $E_1 = E_2 = hv$, 126
    $Fc = mv^2/r$, 40
    $F = KM_1M_2/d^2$, 45, 48
    $F = KQ_1Q_2/d^2$, 95
    $hv = w + \frac{1}{2} mv^2$, 129
    $hv_{max} = Ve$, 128
    $i = v/r$, 100
    $KE = \frac{1}{2}mv^2$, 59
    $l = l_o\sqrt{1 - v^2/c^2}$, 119
    $\frac{1}{2}mv^2 = ev$, 123
    $m = m_o/\sqrt{1 - v^2/c^2}$, 119
    power = work/time, 53
    pressure × volume = constant, 84
    pressure/temperature = constant, 88
    pressure = force/area, 65
    $PE = m \cdot g \cdot h$, 57
    $R_t = R_1 + R_2 + R_3 + \ldots$, 105
    $1/R_t = 1/R_1 + 1/R_2 + 1/R_3 + \ldots$, 103
    $t = t_o/\sqrt{1 - V^2/C^2}$, 119
    $T_c = 5/9 \, (T_f - 32)$, 80
    $T_f = 9/5 \, T_c + 32$, 80
    Torque = Force × length, 36
    $V = d/t$, 1
    $V = g \cdot t$, 11
    volume = length × width × height, 69

Index 139

Equations (con't)
 volume/temperature = constant, 86
 $v = F$, 116
 $(V_1)^2 + (V_1)^2 =$ hypotenuses, 33
 work = force·distance, 49

Falling bodies, 10–16
Floating, 71
Force
 action, 27
 centripetal, 39
 frictional, 19
 gravitational, 48
 moment of, 35
 reaction, 27
Frequency, 74, 116
Friction, 34
 air, 10

Galileo, 12
Gamma rays, 131, 134
Gases, 79
Gas law, Ideal, 89
Gay-Lussacs' Law, 88
Gravity, 10, 44, 57, 93, 95
 acceleration, 11, 15, 22
 constant, 11, 45, 48
 units, 11
Grid, 30
Ground state, 125

Half-life, 131
Heat, 82
Height, 57
Horizontal component, 33
Horse power, 53
Hypotenuse, 32

Illumination, 113
Inertia, 18
Inversely proportional, 85, 95
 "inversely proportional to the square of the distance", 45

Kilocalories, 82
Kilowatt-hours, 108
Kilowatts, 108
Kinetic energy, 124

Latent heat, 81
 ice, 82
 water, 83
Lever arm, 35
Light, 113
 illumination, 113
 intensity, 113
 red, 76
 velocity, 77, 116, 118
Liquid, 79
Lorentz, 118
Lumens, 113

Mass, 18, 25, 26, 46, 47, 57
 rest, 119
Mathematical manipulation, 41, 42, 51, 61, 62, 68, 98
Michelson, 116
Millikan, 123
Moment of force, 35
Motion, 59
 circular, 39
Music, 76

Newton, 17, 33, 39, 43, 45, 93
 first law, 17
 second law, 18
Nucleus, 133

Ohm's Law, 99, 100
Ordered pair, 31
Origin, 30

Parallel resistance, 102
Photoelectric effect, 129
Photon, 129
Piano, 76
Pivot, 35, 36
Plancks' Constant, 126
Planetary system, 125
Pound, 43
Power, 53, 106
Powers of ten notation, 76, 97
Pressure, 65, 68
Problem solving, 2
 equations, 3, 4, 12
 format, 2
 given — find, 12
 substitution, 2

Index

"Product of the Masses", 45
Proportionality constant, 44
Protons, 93, 127, 133
Pythagorean Theorem, 32

Radical, 42
Radioactivity, 130, 133
Radio waves, 117
Radius, 40
Reference level, 57
Relativity, 118, 120
  special theory, 118
Repulsion, 95
Resultant, 30, 33, 34
Resistance, 100
  resistor, 103
Rutherford, 133

Schematic diagram, 100
Scientific notation, 76, 97
Series resistance, 102
Sine wave, 73
Solids, 79
Speed
  definition, 1
  sound, 75
Square root, 42
Systems of measure
  British, 13, 18, 40
  metric, 13, 18, 40
  table, 64

Teeter-totter, 35
Television, 116
Temperature, 79
  Celsius, 79
  centigrade, 79
  conversion, 80
  fahrenheit, 79

Kelvin, 79
Rankine, 79
Thomson, 123
Tide, 48
Time, 5
  dilation, 119
Torque, 35, 36, 37
Transmutation, artificial, 135

Universal Law of Gravitation, 43, 48

Vectors
  diagram, 32
  direction, 29
  horizontal component, 33
  magnitude, 29
  resolution into components, 50
  solution by ordered pairs, 31
  solution by parallelogram, 30, 33
  vertical component, 33
Velocity, 1
  definition, 1
  final, 6, 7
  initial, 6, 7, 14
  units, 1, 5
Vertex, 30
Volt, 99
Volume, 69

Watt, 53, 106
Waves, 11
  electromagnetic, 16
  length, 74, 116
  velocity, 74
Weight, 19, 22, 23
Work, 49
  as a vector, 50

X-rays production, 127